Fundamentals of Electricity

Volume 1
Basic Principles

CONSUMERS POWER COMPANY

Programed by

R.J. CLEAVER, E.J. MEEUSEN, AND R.A. WELLS, Jr.

ADDISON-WESLEY PUBLISHING COMPANY
Reading, Massachusetts
Menlo Park, California · London · Amsterdam · Don Mills, Ontario · Sydney

ISBN 0-201-01185-9
QRSTUVWXY-AL-898765432

Seventeenth Printing, April 1982

ADDISON-WESLEY PUBLISHING COMPANY, INC.
Reading, Massachusetts · Palo Alto · London
New York · Dallas · Atlanta · Barrington, Illinois

ADDISON-WESLEY (CANADA) LIMITED
Don Mills, Ontario

Preface

This course of study has been developed by employees of Consumers Power Company, an investor-owned utility serving much of the State of Michigan. While it was originally prepared for use by Company employees to assist them toward a better understanding of their job and the nature of the energy served their customers, it has been revised to pertain to a more general audience. The program has been tested by approximately 500 employees, including clerks, linemen, appliance servicemen, electric metermen and engineering technicians. Their educational backgrounds varied all the way from the eighth grade to two to four years in college. The average score of the entire group showed an average increase from 53 percent on a pre-test to 87 percent on a similar post-test, after completion of the program.

The authors are indebted to many for valuable assistance and advice in the preparation and revision of this program. Among those who were most helpful were: Mr. C.G. Valentine, Michigan Bell Telephone Company; Mr. Geary Rummler, Center for Programmed Learning for Business and Industry, University of Michigan; and Dr. William G. Bickert, Assistant Professor, Agricultural Engineering, Michigan State University. Special mention should also go to Mr. E.H. Luther, former General Personnel Development Supervisor, Consumers Power Company, for providing the incentive and stimulation to complete this program.

R.J.C.
E.J.M.
R.A.W., Jr.

Jackson, Michigan
March, 1966

Foreword

TO THE
STUDENT

You will find programed instruction a new and somewhat revolutionary way of learning. Because of being continually involved in the program through your writing a response to a question or statement, you will develop the feeling of very active participation in this learning process. You will also appreciate the opportunity of being able to proceed at your own pace rather than be forced to move more slowly or faster with a group, depending on its pace.

Programed learning or instruction also provides you with the opportunity for immediate feedback to your learning progress. Thus, you answer the question, "How am I doing?" immediately because you are given the correct answer (response) to the question immediately after you've answered it. You no longer must wait a week until the instructor returns your graded exam to learn where you stand. In this way, too, your learning is always based on positive results. You know you are on the right track and you can proceed with more confidence.

HOW DOES
PROGRAMED
LEARNING
WORK?

In a course using programed instruction as a teaching technique, you are given information in different-sized quantities from a short sentence to a longer paragraph. This quantity of information is called a frame and within this frame is usually a question to answer or a statement to complete, based on information given in the frame. You will be asked to answer this question, and immediately upon completion of your response, you will be given the correct answer, which you should compare with your own constructed response. (Where there is a choice of bracketed words, cross out the answer you do not want.) The answer to a frame is found to the right of it. Answers separated by slashes indicate alternatives.

If you are correct, continue on to the next frame where you will repeat the process. Should your answer be incorrect, either you will be given information that tells you "why" the given answer was incorrect or you may review the previous sequence of frames to help you discover your error.

It is important that you write your answer in the space provided. This will help you remember the information given to you. It is also important that you answer the question in the frame before looking at the correct answer. At the back of the book, you will find a response mask. Use it to cover the correct response until you are ready to compare it with your answer. No one will be "grading" your work in the programed textbook — it is your own to use as you desire during and after the course. Therefore, while you could easily look ahead on every frame, you would only be cheating yourself!

At times, you may find frames which do not require a response. These are informational or "lead-in" frames which are necessary to introduce a new concept or idea. You will also notice within the program references to panels (for example, II-D, III-A, V-C, etc.). These are illustrations or drawings intended to make the information in that frame, or sequence of frames, easier for you to understand. They appear in the perforated section at the end of this book.

Fundamentals of Electricity, Volume 1: Basic Principles, is a simply written program designed to develop an understanding of certain basic electrical fundamentals. Upon completion of this course, the student should

ABOUT THIS COURSE

1. be familiar with basic electrical terms and symbols,

2. understand fundamental electrical laws and the application of these laws,

3. have a general knowledge of the nature of electricity and how it is generated, transmitted, and used.

The objective of this program is purely educational; it is not intended that the information be used as standards for operation or construction. Some areas of the program have been written more for ease of learning and logical thought than for technical accuracy.

The course has been organized into eight learning blocks (chapters) for better presentation and learning. They are the following.

    I. Electricity — electron flow
   II. Ohm's law
  III. Series circuits
  IV. Parallel circuits
   V. Power
  VI. High and low voltage
 VII. Magnetism (generators and transformers)
VIII. Summary (practical applications)

In addition, you will find a glossary of electrical terms and an alphabetized topic index at the back of your program. These are for reference during use of the program and after completion of it. As you work through the program, use only the right-hand pages until you reach Frame 65, page 79, then reverse the book and continue with Frame 66, page 80, again using only the right-hand pages until you reach the end of the book.

The course is intended for individualized instruction. Thus, you should proceed through the program at your own pace. Most students will take about 20 hours to complete the program but this overall time will vary with students, depending on their reading ability, previous experience in electricity, and their mathematical ability. So don't be concerned if others seem to be moving through the program at a more rapid pace.

Also, work through the program for only as long a period of time as you can effectively learn. When you become tired, either lay the program aside until the next session or at least take a five- or ten-minute break. Most students find that an hour of study in programed instruction at one time is sufficient. You will find that you will enjoy this program more and your learning will be more effective if you follow these few simple suggestions:

1. Read each frame carefully. Be sure you understand completely before proceeding to the next frame.

2. Write your answer to each frame.

3. Avoid looking at the correct response until you have written your own answer. Use the mask provided to cover the correct response.

4. Work at your own pace.

5. Take a break when you become tired.

Good luck!

Contents

Pre-test

Answer the following questions to the best of your ability. Use as many or as few words as necessary. You will find some questions very easy, some more difficult. You are not expected to be able to answer all of them correctly.

1. _____ is the unit of measure for the force which causes electric current to flow through resistance in an electric circuit.

2. The resistance in an electric circuit is measured in _____.

3. The current flowing in the circuit is measured in _____.

4. What is the formula which expresses the relationship between voltage, current, and resistance in an electric circuit? _____.

5. A _____ is a protective device installed in a circuit to protect against overloading the circuit.

6. Power in an electric circuit is measured in _____.

7. The term kilo, sometimes used with electrical terms, means _____. A kilowatt, therefore, is equal to _____ watts.

8. The loads in an electric circuit can be connected in two ways: _____ or _____.

9. If you had two conductors which were both the same length but one had twice the diameter of the other, which would carry the most current? [Larger/Smaller.] Which would have the most resistance? [Larger/Smaller.]

10. Will a 120-volt circuit carrying a 2200-watt load blow a 15-amp fuse? [Yes/No.] Why? _____

11. The formula for the power requirement of an electrical device is ___ = _____.

12. The normal circuit wiring in a house is an example of a [series/parallel] circuit.

13. A current of _____ amp will flow in the circuit below.

120 volts   30 ohms

14. What is the principal advantage of a three-wire entrance service over a two-wire service? _____

_____.

15. A No. 14 conductor [will/will not] safely carry a 100-amp load. Why? _____

16. A current of ___ amp will flow in the circuit below.

17. A [copper/glass] rod is the best conductor of electricity.

18. A circuit carrying 4 amp of current flowing through a 25-ohm resistance develops ____ watts of power.

19. In every electric circuit, the current flows from the [positive/negative] terminal to the [positive/negative] terminal.

20. A transformer has 20 turns in the primary windings and 5 turns in the secondary windings. If the primary voltage is 100 volts, the secondary voltage will be ____ volts.

21. The transformer described in Problem 20 would be a [step-up/step-down] transformer.

22. A load that has 30-ohm resistance and is rated at 750 watts would have ___ amp of current flowing through it.

23. At 2 cents per kwh, how much does it cost to operate a 1000-watt iron for $\frac{1}{2}$ hour? ____

24. With a voltage generated at 60 cycles per second, how many times will the electricity be "off" in a five-second interval? ____

25. The total resistance in the circuit below is ___ ohms.

26. Current [will/will not] flow in the circuit below. Why? _____

_____

27. In some electrical appliances, for example washers and dryers, a wire is often connected from the case of the appliance to a

cold-water pipe. This wire is called a _____ wire. What is its purpose? _____

28. If an additional 2-ohm parallel resistance were added to the circuit in Problem 25, the total current for the circuit would [increase/decrease].

29. Current flowing in a conductor causes a certain amount of _____ _____, which may result in low voltage being supplied to the load.

30. What happens to a wire that is forced to carry an amount of current in excess of its rated capacity? _____

_____

31. A motor rated at 120 volts, when connected to a 240-volt outlet, would _____.

32. If you noticed in your home that your lights were sometimes dim, motors were running slowly, etc., you would assume that you had a condition of _____ _____.

33. In some modern electric water heaters, 240 volts are supplied to a 4800-watt element. How much current will this element draw? _____ amp.

Answers to Pre-test

1. Voltage

2. ohms

3. amp (amperes)

4. $I = \dfrac{E}{R}$ (E = IR or $R = \dfrac{E}{I}$)

5. fuse

6. watts

7. 1000
   1000

8. series
   parallel

9. Larger.
   Smaller.

10. Yes.
    2200 watts at 120 volts =
    18.3 amperes

11. $P = I^2 R$ or $P = IE \times PF$

12. parallel

13. 4

14. 240 volts are available only
    from a three-wire service.
    A three-wire service allows
    a better load balance.

15. will not
    A No. 14 wire is nominally
    rated to carry 15 amperes
    safely. It would get very hot
    if forced to carry 100 amp-
    eres.

16. 8

17. copper

18. 400

19. positive
    negative

20. 25

21. step-down

22. 5

23. 1 cent.

24. 600 (120 times per second)

25. 2

26. will not
    Switch is open.

27. ground
    To prevent electric shock
    in event of equipment mal-
    function.

28. increase

29. voltage drop

30. It will overheat, cause volt-
    age drop, and if the current
    is extremely high, the wire
    may melt or cause a fire.

31. burn out
    (draw excessive current)

32. low voltage

33. 20

I Electron Flow

[1] In order to understand electricity, we must first study certain fundamental laws of physics. Physics tells us that basically electricity is the movement of forms of matter. So, to begin our study of electricity, then, we must first understand what is meant by the term <u>matter</u>.	**[1]**
[2] <u>Matter</u> is defined as anything that has weight and takes up space. Then anything that you can feel, touch, or see is considered _____.	**[2]** matter
[3] Water, wood, iron, air, and salt are examples of _____. Why?_____ <div align="center">(Own words)</div>	**[3]** matter Because they all have weight and take up space, or because you can see and feel all of them.
[4] Matter can be subdivided into smaller particles called <u>molecules</u>. Water, which is matter, consists of a very large number of _____ grouped together.	**[4]** molecules
[5] Molecules grouped together make up _____.	**[5]** matter

[6]

Molecules consist of two or more smaller particles called <u>atoms</u>. Since the water molecule is made up of two types of small particles, called hydrogen and oxygen, the hydrogen and oxygen particles are called _____.

[6]

atoms

---

[7]

To summarize what we have learned so far, everything around us is composed of _____. We can subdivide this matter first into _____ and then again into _____.

[7]

matter

molecules

atoms

---

[8]

From the study of the atom, it was found that the atom, too, is made up of even smaller units, which are called "electrons," "protons," and "neutrons." Thus the atom can be composed of at least _____ kinds of particles.
    (Number)

[8]

3

---

[9]

The center of the atom is called the mass or <u>nucleus</u> of the atom, and around the atom are imaginary <u>rings</u>. Imaginary _____ encircle the n_____ of the atom. (See Panel I-A.)

[9]

rings

nucleus

---

[10]

The mass or nucleus contains the protons and neutrons. The rings around the nucleus contain the e_____. (See Panel I-A.)

[10]

electrons

Conclusion

Now that you have completed this course in basic electricity (which we hope will help you on your job), as a final step, refer to Panel VIII-A and follow through, step by step, the procedure in serving the customer with electric power:

1. The electricity is first mechanically generated at the generating plant.

2. Immediately outside the generating plant the voltage from the generator is connected to a step-up transformer. This transformer increases the voltage to a level at which it can efficiently be transmitted over the transmission line.

3. The power from this location is then carried over the transmission line (maybe many miles) to a transmission substation.

4. At this transmission substation the voltage is reduced by a step-down transformer to a subtransmission voltage, which is still greater than distribution voltage. This subtransmission voltage serves large industrial and commercial customers who have their own transformers to reduce it to the value they require.

5. This subtransmission voltage is used also to supply the distribution substation with power.

6. The voltage is reduced by a step-down transformer at the distribution substation to the distribution primary voltage.

7. The distribution primary lines carry this voltage to residential and rural areas.

8. The primary distribution voltage is then reduced to the 120/240-volt secondary distribution voltage which serves homes.

9. This final reduction in voltage occurs at the step-down transformers mounted on the distribution poles throughout these areas.

10. This secondary distribution voltage enters the homes through the service drop and house entrance.

[11]

Now that we have pictured the atom and its components, let's leave it for a moment and define an <u>electric charge</u>. An electric charge is a particle which has either a positive or a negative charge. Therefore, a particle with a negative charge is an example of an _____ _____.

[11]

electric charge

---

[12]

In the electrical world, there are symbols or signs to denote the two charges. A positive charge is denoted by + and a negative charge is denoted by −. A _____ denotes a posi-
(Draw sign or symbol)
tive charge and _____ denotes a negative charge.
(Draw sign or symbol)

[12]

+

−

---

[13]

A charge that has the symbol or sign − is _____.

[13]

negative

---

[14]

A _____ sign denotes a positive charge.

[14]

+

---

[15]

The two charges that we will work with are either _____ or _____.

[15]

positive / +

negative / −

[104]

Please remember that this has been only a beginning course in basic electricity. It was intended to familiarize you with electrical terms and fundamental laws, no more. There are many, many more electrical concepts with which you should perhaps become familiar in order to better perform your job. These remaining ideas may be presented in future courses of this nature but, in general, you should now have sufficient basic knowledge to study these additional principles on your own.

[104]

[16]

A charged particle is either _____ or _____.

[16]

positive

negative

---

[17]

If a certain particle carries a positive (+) charge and another particle carries a negative (−) charge, they are said to be opposite in electric charge. A positive charge is the _____ of a negative charge.

[17]

opposite

---

[18]

Scientists learned that opposite electric charges <u>attract</u> each other. Therefore, a positively charged particle would _____ a negatively charged particle.

[18]

attract

---

[19]

For example, a positively charged piece of iron would be attracted to a _____ charged iron ball. (See Panel I-B.)

[19]

negatively

---

[20]

Since it has been shown that oppositely charged particles attract each other, it can also be seen that similarly charged particles will <u>repel</u> each other. Thus, a positively charged particle will _____ another positvely charged particle.

[20]

repel

[99]

The system neutral ground, therefore, is not designed to prevent electric shock as with an appliance ground. It is to be used to electrically connect all of the neutral wires in the system and to carry part of the system neutral _____.

[99]

current

[100]

An appliance ground does not carry _____, whereas the system neutral ground carries a small amount of _____.

[100]

current

current

[101]

The appliance ground protects the user against electric _____ by causing the circuit _____ to blow when electrical conditions become abnormal.

[101]

shock

fuse

[102]

The system neutral ground is not intended primarily as a protection against _____ _____ as is an appliance ground.

[102]

electric shock

[103]

You have now studied several of the basic individual principles of electricity as they apply to your job. This final section has attempted to put together these basic concepts to show the type of electrical systems that are required to transmit the power from generating plants to customers for their use. It would be wise for you to review all of the panels in this section now and periodically in the future, for they will summarize all of the individual concepts we have discussed in this course.

[103]

[21]

Thus, two negatively charged particles will be found to _____ each other. (See Panel I-C.)

[21]

repel

---

[22]

As another example (see Panel I-C), a positively charged iron ball will repel another iron ball with a _____ charge.
                              (Sign)

[22]

+

---

[23]

Think back to the previous discussion of the atom, where we learned that each atom is composed of three particles, neutrons, protons, and _____.

[23]

electrons

---

[24]

In the atom, the electrons are found in the imaginary rings, while the nucleus contains the neutrons and _____.

[24]

protons

---

[25]

The neutron is the name given to a particle with no electric charge, while a proton is the name of a positively charged particle. Thus, the electron, which is the opposite of proton, is the name given to a particle with a _____ charge.
                              (Sign)

[25]

−

[94]

(3) Whatever current leaves the transformer over one of the "hot" wires, after flowing through the connected loads, must <u>all</u> return to the transformer winding and in this case to the neutral connection of the winding.

[94]

[95]

(4) Part of the neutral current flowing through the light bulb returns to the transformer through earth, whereas the majority of it returns through the neutral wire.

[95]

[96]

Refer to Panel VIII-L. The basic difference between an appliance ground and a system neutral ground is that there will normally be a small amount of current flowing in a neutral ground. Since earth is a conductor and offers an alternate path, it can be seen that this "ground conductor" is in [series/parallel] with the system neutral wire.

[96]

parallel

[97]

Since the "ground conductor" is in parallel with the system neutral, there are therefore _____ paths through which the neutral current can flow.

[97]

two

[98]

Therefore, the system neutral ground serves to connect all of the neutral wires in the system and establishes a [parallel/series] path for the flow of neutral _____.

[98]

parallel

current

[26] Therefore, the atom contains two kinds of electrically charged particles: the proton, which has a _____ charge, and the _____, which has a − charge. (Refer to Panel I-D.)	[26] positive electron
[27] Since the nucleus of the atom contains protons, it is said to have a _____ charge. (Sign)	[27] +
[28] Since the electrons in the imaginary rings around the atom are negatively charged particles, they will be [attracted/repelled] by the nucleus.	[28] attracted
[29] The electrons in the imaginary rings are constantly spinning about the nucleus of the atom and are therefore trying to get farther away from the nucleus. This can be likened to the boy rotating a ball at the end of a string (see Panel I-E, Part a). The ball pulls away from the boy. The ball does not fly completely away from the boy because of the _____ attached to the ball and held by the boy.	[29] string
[30] As the string holds the ball in place, the <u>attraction</u> between the negative electrons and the positive protons in the nucleus holds the electrons in their imaginary rings. If this _____ were not there, the electrons would fly completely away from the _____. (See Panel I-E, Part b.)	[30] attraction nucleus / protons

[89]

It is necessary to have a neutral wire in a service entrance, therefore, to obtain a voltage of _____ volts.

[89]

120

---

[90]

Refer to Panel VIII-K. Since there are appliances connected to this neutral, there will be a certain amount of current flowing in the _____ wire.

[90]

neutral

---

[91]

Refer to Panel VIII-K. The system neutral is physically connected to ground. (One method is by ground rod, a pipe driven into the earth on one end and fastened to the wire on the other end.) The system neutral, therefore, has the same electrical potential as the _____.

[91]

ground / earth

---

[92]

Refer to Panel VIII-K. The voltages existing between the system neutral and the "hot" wires from the transformer are shown in Panel VIII-L. The hot wires have a voltage of 240 volts between them. The neutral wire is then used to obtain _____ volts from either "hot" wire.

[92]

120

---

[93]

From Panel VIII-K note the following through Frame 95:

   (1) 120 volts are obtained from the transformer by connection of a neutral wire to the center of the winding of the secondary.

   (2) Both the neutral of the transformer and the neutral bar of the fuse box are connected to ground by grounding rods, water pipes, etc.

[93]

[31]

The strength of this attraction depends on the distance between the electrons and the nucleus. The greater the distance of the electron from the nucleus, the weaker the attraction will be. Thus, to have a strong attraction, the electrons must be _____ to the nucleus.

[31]

close / closer / nearer

---

[32]

Referring to Panel I-D, the electrons in Ring 3 will have a [stronger/weaker] attraction to the nucleus than will the electrons in Ring 1.

[32]

weaker

---

[33]

An atom in which the attraction between the nucleus and the electrons is weak is said to have <u>free</u> electrons. Then a material made up of millions of atoms with a weak attraction between the electrons and the nucleus will have many _____ electrons.

[33]

free

---

[34]

If an atom has the same number of protons (positive charge) as electrons (negative charge), the atom itself is said to be <u>neutral</u>. For example, an atom with four electrons and _____ protons is said to be neutral.

[34]

four

---

[35]

An atom with seven protons in the nucleus and seven electrons in the outer rings is said to have a _____ charge.

[35]

neutral

[84]

The reference to a system neutral ground is somewhat different from that to an appliance ground. It will not be possible to fully discuss system neutral ground without a knowledge of transformer theory. However, since you will hear this term, there are certain concepts you should become familiar with.

[85]

neutral

System neutral, in the first place, refers to the neutral wire leading to the house from each transformer. It is called system neutral merely because it exists everywhere on the system. We call it neutral since it has no charge with respect to ground. Under normal conditions, then, no voltage exists between ground and the system _____ wire.

[86]

Why do we have a system neutral? We saw before that we can get 120 volts from a 240-volt transformer by attaching a wire to the center of the winding as shown below.

[87]

240
120

This wire is called a neutral wire because it is in a neutral position (or in the middle) with reference to the ends of the secondary winding. From the ends of the winding we would measure, on a 120/240-volt transformer, _____ volts. From the neutral wire to either end of the winding we would measure _____ volts.

[88]

120

Refer to Panel VIII-K. From neutral to a "hot" wire we have a voltage of _____ volts, which is the voltage required to light a bulb, heat a toaster, etc.

[36]

If we take a neutral atom and add an electron, we say that this atom is in excess of electrons. Therefore, an atom in excess of electrons will have _____ electrons than protons.

[36]

more

[37]

An atom with an excess of electrons is no longer neutral. The excess of the negative electrons will give the atom a _____ charge. (See Panel I-D.)

[37]

negative

[38]

By the same token, if an electron is removed from a neutral atom, this atom is lacking or deficient in electrons. If an electron is removed from a neutral atom, the atom has more _____ than electrons.

[38]

protons

[39]

An atom that is lacking in electrons will have more protons than electrons. An atom with more protons than electrons has a _____ charge.

[39]

positive

[40]

The number of protons in an atom never changes since only the electrons are able to escape the atom. Therefore, the charge that the atom acquires, either positive or negative, depends on whether or not it has an excess or deficiency of _____.

[40]

electrons

[79]
Another method of grounding an appliance is to run a grounding wire along the house circuitry. This grounding wire serves the same purpose as the earth. Instead of the excess current flowing through the earth, it will now flow through the _____. (See Panel VIII-J.)

[79] grounding wire

[80]
The appliance case is connected to the ground wire through a three-prong plug. If an electrical short should occur to the case of the appliance, the excess current will flow through the _____, which will in turn cause the fuse to blow or burn out.

[80] ground wire

[81]
Now we have two methods of grounding an appliance: first we could connect the appliance to a cold-water pipe, which will allow us to use the _____ as the path for excess current, or we could run a _____ in the house circuitry and connect the appliance to this by use of a three-prong plug.

[81] earth / ground wire

[82]
Regardless of which method is used to ground the appliance, the purpose of grounding is to prevent _____ _____ to the user of the appliance.

[82] electric shock

[83]
An appliance ground, therefore, is used to prevent electric _____ and under normal conditions carries no _____.

[83] shock / current

[41] For example, if there is an atom with eight protons in the nucleus and seven electrons in the outer rings, it will have a _____ charge. If the atom has nine electrons in the outer rings, it will have a _____ charge.	**[41]** + (positive) − (negative)
[42] A storm cloud contains atoms with an excess of electrons and, therefore, it has a _____ charge. (See Panel I-F.)	**[42]** negative (−)
[43] The earth, by its make-up, contains atoms deficient in electrons and, therefore, it has a _____ charge. (See Panel I-F.)	**[43]** positive (+)
[44] We have seen from previous examples that unlike charges attract each other. Therefore, there is an attraction between the storm cloud and the earth just mentioned. This attraction exists because the cloud and the earth have _____ charges. (See Panel I-F.)	**[44]** opposite / unlike
[45] This attraction between the cloud and earth is called an <u>electric force</u>. We say, therefore, that an _____ _____ exists between unlike charges.	**[45]** electric force

[74]

Therefore, if an appliance is properly _____, the danger _____ of electric shock to the user is removed.

[74] grounded

[75]

Under normal conditions, no current flows through this appliance ground path. If the bare wires from the entrance panel touch the case of the appliance, high _____ will flow, causing the fuse to _____ and removing the danger of electric _____.

[75] current / melt / blow / shock

[76]

Therefore, by grounding both the system neutral from the entrance panel and the _____ of the washer, or any other electrical appliance being used, we can protect the user against electric _____.

[76] case / shock

[77]

Appliance grounds are therefore used to prevent _____ and do not normally conduct _____.

[77] electric shock / current / electricity

[78]

Refer to Panel VIII-1. The previous method of appliance grounding was to attach the neutral from the entrance panel and the appliance case to ground through a cold-water pipe or to use some similar method. By doing this we used earth as a c_____.

[78] conductor

[46] This electric force between the cloud and the earth causes the electrons to flow from the source of excess electrons to the source of deficiency of electrons or from − to ____.	[46] +
[47] In a storm this flow of electrons from the negative (excess) to the positive (deficient) source takes the form of lightning. (See Panel I-G.) Lightning will continue as long as the electric _____ continues.	[47] force / potential / difference
[48] The electric force between the cloud and the earth will continue until the cloud and the earth have a balance in the number of electrons and have neutral charges. Electrons will not flow between two sources of _____ charge. (See Panel I-H.)	[48] neutral / like / same / equal
[49] As the attraction becomes greater between a negative (−) charge and positive (+) charge, we could also say there is an increase in the e_____ f_____.	[49] electric force
[50] As this electric force becomes larger in value, it causes <u>more</u> electrons to flow from the negative point to the positive point. Higher electric forces cause _____ electrons to flow.	[50] more

[69]

See Panel VIII-1, Part (b). When the fuse blows, the circuit is no longer complete and there is no longer any _____ on the case of the washer; therefore, no shock hazard exists.

[69] voltage

[70]

Panel VIII-1 showed a grounding condition where the appliance case was connected by wire to a cold-water pipe. This water pipe, if in a city water system, will run into the earth. Thus, the appliance has a metallic path to earth, or we say it is _____ed.

[70] grounded

[71]

See Panel VIII-1. Also note from this panel that the neutral wire from the system is also connected to ground by the same method. Therefore, the system neutral wire in the panel is also _____.

[71] grounded

[72]

The appliance case and the system neutral are, therefore, connected by a low resistance through the earth. In the event that the wire from the fuse touches the appliance case, the current will no longer flow through the resistance of the motor but will flow through the low-resistance _____ circuit.

[72] earth / ground

[73]

Since this earth-circuit resistance is very, very low, _____ current will flow, causing the _____ to melt and removing the _____ from the case of the appliance.

[73] high fuse voltage

[51]

We have stated that an electric force exists when two charges are different, namely, positive and negative. To repeat, electric force exists when there is a _____ between two charges.

[51]

difference

---

[52]

We saw previously that electric force caused the flow of electrons between points of unlike or _____ charges.

[52]

different / opposite

---

[53]

We can also say that this electric force has the ability or potential to cause these electrons to flow. The number of electrons that will flow depends on the amount of _____, or electric force.

[53]

potential

---

[54]

In electrical language, therefore, we call this electric force a difference of potential. The greater the _____ ____ _____, the greater the number of electrons that will flow.

[54]

difference of potential

---

[55]

When electrons are in a state of imbalance of charges between two points, we have a _____ of _____.

[55]

difference

potential

[64] We will now see why this ground is a safety feature. When we say ground we mean: run a wire from the appliance to the earth through any convenient metallic conductor. A conductor which connects an appliance or load to earth is a _____ conductor.

[64] ground

[65] We saw earlier that it is possible to use the earth as a conductor since the earth or ground will conduct _____. We will now make use of this fact.

[65] current / electricity

[66] Refer to Panel VIII-1, Part (a). Under normal conditions of operation the resistance of the motor keeps the current in the circuit at a low enough value to prevent the fuse from melting. If we were to greatly reduce this resistance, the current would greatly _____ and the fuse would _____.

[66] increase
blow

[67] Panel VIII-1, Part (b) shows what would happen if the wire leading from the fuse were to touch the case of a washer. The current will try to follow the path with the least resistance. The path with the motor in it has 12 ohms; the path through the case and through ground has a much smaller resistance. Through which circuit will more current flow? [Motor circuit/Ground circuit.]

[67] Ground circuit

[68] Refer to Panel VIII-1, Part (b). The current will flow through the ground circuit, which contains very little resistance. This will allow a _____ amount of current to flow. Since the fuse is rated at only 15 amp, this abnormally high current will cause the fuse to _____.

[68] large / great
melt / blow

[56]

The electrical world needed a unit of measure for this potential difference. So, in honor of Mr. Volta, who did a lot of experimentation with electricity, they called this unit of measure the volt. The electric force or potential difference that causes the flow of electrons is measured in _____ .

[56]

volts

[57]

To push water through a hose there must be pressure or force. To cause electrons to move there must also be force. This force or potential is referred to as _____ .

[57]

volts / voltage

[58]

Instead of having 120 potential difference in your house wiring, the more common name would be 120 _____ .

[58]

volts

[59]

Just as nature created a potential difference, man can also create it. Through chemical reaction, use of heat, magnetism and friction, man can create a potential difference. Since man created this _____ _____ , he can use it to his advantage.

[59]

potential difference

[60]

Let's take a look at a potential difference which man has created. This means that there will be a point of excess in electrons (−) and a point of deficiency in electrons (+). Now, connect these two points with a wire containing many free electrons. (Remember that free electrons can move easily.) These free electrons will be [attracted/repelled] to the + point. (See Panel I-I.)

[60]

attracted

[59]

Panel VIII-H shows an entrance panel and typical house circuits. List the voltages of the following circuits.

(1) Circuit 1 _____

(2) Circuit 2 _____

(3) Circuit 3 _____

[59]

(1) 120 volts

(2) 120 volts

(3) 240 volts

---

[60]

Refer to Panel VIII-H. In which circuit would you connect the following loads to avoid overloading one circuit as much as possible?

(1) 240-volt electric range _____   (4) 120-volt mixer _____

(2) 240-volt water heater _____   (5) 120-volt iron _____

(3) 120-volt toaster _____   (6) 120-volt frypan _____

[60]

(1) 3   (4) 1, 2

(2) 3   (5) 2, 1

(3) 1, 2   (6) 2, 1

[For (3)-(6) answer must not show all loads connected to one circuit.]

---

[61]

We should mention "grounding." This can be interpreted in two ways. One form would be system neutral grounding and the other form is appliance grounding. We will take each form separately.

[61]

---

[62]

Refer to Panel VIII-I, Part (a). Notice that the wires furnishing the electricity to the washer are insulated and do not touch the case of the washer. So long as this condition exists, there is no problem. But if the wire from the fuse gets pinched or cut and some of its insulation is scraped off, it can touch the case of the washer, thus giving it a voltage.

[62]

---

[63]

This voltage could cause electric shock to the user. To prevent this, we ground the washer (or similar appliance). This _____ will prevent the user from receiving an electric shock.

[63]

ground

[61]

As the wire "lets go of" its free electrons to the + point, the wire becomes deficient in electrons. When the wire is deficient in electrons, the excess electrons at the negative point will be attracted by the wire, which becomes _____ as it loses its electrons. (See Panel I-I.)

[61]

+ / positive

[62]

The process will continue with the + point attracting the free electrons in the wire, causing the wire to be deficient in electrons. The wire will then attract the free electrons from the – point. This process will continue until the electrons have balanced out. Or if we keep an excess of electrons at the – point and if the + point is kept deficient of electrons, then this process will last indefinitely. (See Panel I-I.)

[62]

[63]

This flow of electrons from negative to positive is called electric current. Electron flow is known as _____ _____

[63]

electric current

[64]

Again, we need a unit of measure for the current flow in a conductor. The amount of current flowing in a conductor is measured in amperes. When measuring the current flow, we will use the term _____. [Note: While the term "current flow" may not be technically correct, this is common terminology and hence will be used for clarity in this program.]

[64]

amperes

[65]

An ampere (commonly abbreviated "amp") is a measure of the number of electrons flowing in a conductor. Therefore, as the electric current in a conductor increases, we will say that the number of _____ of current increases.

[65]

amperes

[54]
The unit of measure for the energy we use is the _____.
This means that _____.
(Own words)

[54]
kilowatt / watt-hour
we have used power for a
certain length of time.

[55]
From the electric meter, the service drop continues into the customer's entrance panel or fuse box. The entrance panel is the place where the house wiring circuits connect to the service drop from the system of the utility company. Fuses which protect the customer's house wiring are also located in this _____ .

[55]
entrance panel / fuse box

[56]
The voltages from the secondary of the distribution transformer are available to the customer in his entrance panel. The customer therefore has _____ / _____ volts available if he has a _____ - wire service drop.

[56]
120/240
three

[57]
Refer to Panel VIII-G. What voltage would exist in the customer's entrance panel between terminals
(1) AB? _____
(2) BC? _____
(3) AC? _____

[57]
(1) 120 volts
(2) 120 volts
(3) 240 volts

[58]
Referring again to Panel VIII-G, indicate the terminals across which you would connect the following devices.
(1) 240-volt dryer _____
(2) 120-volt bulb _____
(3) 120-volt toaster _____
(4) 240-volt range _____

[58]
(1) AC
(2) AB, BC
(3) AB, BC
(4) AC

[66]

When the potential or voltage causes electron flow, we will call this electron movement _____ _____.

[66]

electric current

[67]

To have electric current, there must be a potential or electric force which we call _____.

[67]

voltage

[68]

Now, if we had connected these + and − points with a material that did not have any "free" electrons, our flow of electrons from − to + could not get started. Or you might say that the material resisted the flow of electrons; you could say that a material without free electrons will _____ the flow of electrons.

[68]

resist

[69]

If we wanted to limit the electron flow or electric current, then we would put a r_____tance between the − and + points.

[69]

resistance

[70]

Rubber, glass, plastic are examples of materials that do not have free electrons. Therefore, the above materials would _____ the flow of electrons.

[70]

resist

[50]

Since you always use power for a certain length of time, even though it may be only one minute, every time you use power you use _____ even though it may be a very small amount.

[50]

energy

---

[51]

Refer to Panel VIII-F. How many watt-hours would be registered by the meter for each circuit.

(a) _____ watt-hours

(b) _____ kwh

[51]

(a) 2200
Explanation: 1000 watts + 100 watts = 1100 watts for 2 hr; 1100 × 2 = 2200 watt-hours.

(b) 14.4 kwh
Explanation: 10 amp × 120 volts = 1200 watts. 20 amp × 120 volts = 2400 watts. Total 3600 watts. 3600 × 4 = 14.4 for hr; kwh.

---

[52]

A kilowatt-hour or watt-hour is a unit of _____ and a unit of measure of the length of _____ we use electric _____.

[52]

energy

time

power

---

[53]

Every time we turn on a light or plug in an appliance we use _____ and _____.

[53]

power

energy

(either order)

[71]

Just as there was a need for units of measure for electric force
(_____) and current flow (_____), it was also important
that a unit of measure be developed for resistance. This was
given the name ohm in honor of George Simon Ohm, one of the
pioneers in the field of electricity. When the resistance in a conductor increases, the number of _____ are increased.

[71]

volts

amperes

ohms

---

[72]

The current in a conductor depends in part on the number of
_____ of resistance in the conductor.

[72]

ohms

---

[73]

Before proceeding, let's be sure that we understand the three
common electrical terms we have just discussed. Refer to
Panel I-J, if necessary, and answer the following questions.
List the common units of measure for the electrical terms below.

    Force   _____

    Resistance _____

    Current   _____

[73]

volts

ohms

amperes

---

[74]

From what we have just learned, complete the following statements, giving first the electrical term and then its common unit.
Electric _____ or _____ cause _____ or _____
to flow through a conductor, and _____ or _____
oppose this flow of current.

[74]

force (or) volts

current (or) amperes

resistance (or) ohms

---

[75]

Previously when we connected a material between a minus (−) and
a plus (+) point to get electron flow started, we used a material
with many free electrons. Or we might say that a material with
many free electrons will easily carry or "conduct" an electric
current. A material with free electrons is a _____
of electric current.

[75]

conductor

[45]

How much energy would a 750-watt iron use if it were operated at this rating for

10 hours? _____

1/2 hour? _____

[45]

7500 watt-hours

Explanation: 750 × 10 = 7500 watt-hours.

375 watt-hours

Explanation: 750 × 1/2 = 375 watt-hours.

[46]

A more common term than watt-hour for energy used by electric utilities is the kilowatt-hour. The word kilo means 1000. Therefore, one kilowatt-hour is exactly the same as 1000 _____-hours.

[46]

watt

[47]

The kilowatt-hour is the unit used to determine your bill each month. If your total electric energy usage for a month was 500,000 watt-hours, this would be _____ kilowatt-hours.

[47]

500

[48]

The abbreviation for kilowatt-hour is kwh or sometimes kwhr. If you examine your electric bill, you will probably find this symbol somewhere on it. If the charge for each kilowatt-hour we use were 2 cents, how much would it cost us to run a 100-watt bulb for 24 hours? _____

[48]

4.8 cents

Explanation: 100 × 24 = 2400 watt-hours or 2.4 kwh; 2.4 × 2 cents = 4.8 cents.

[49]

If you remember that in general power is equal to the relation $P = IE$, you can see that any time current flows through your meter, you use power since the voltage is always there. The more current you use, the more power and, therefore, the more _____ you use.

[49]

energy

[76]

Since, as we have seen previously, a metal wire serves as a path through which electrons will flow, a metal wire is a _____ of electric current.

[76]

conductor

---

[77]

Wire made of copper, aluminum, or iron can be used as a _____ because these metals have many free electrons.

[77]

conductor

---

[78]

There are four principal conditions that determine the ability of a material to conduct or resist the flow of electric current. They are

    (1) type of material (number of free electrons),
    (2) thickness,
    (3) length,
    (4) temperature.

[78]

---

[79]

Let's take a portion of a copper wire with its many free elec-trons. (See Panel I-K.) In wire 1, the electrons can move along without much interference or resistance. But in wire 2, these same electrons have been "squeezed" down into less space, so that they are now rubbing against the outside walls, which will resist their flow. If we want a large amount of electrons to flow along freely, we would use a [large/small] wire.

[79]

large

---

[80]

In a small wire in which the electrons are squeezed together, the electrons constantly bump each other and the side walls, which resist their movement; therefore, a small wire will offer more _____ to their movement than a large wire, where they can flow freely.

[80]

resistance

[40]

We use a large amount of energy when we use power for a very long period of _____.

time

[41]

Remember that we said power is measured in units called watts and for convenience we measure time in hours. Therefore, the unit of measure for electric energy must be a _____-hour.

watt

[42]

A watt-hour is the unit of measure for _____, which itself measures the amount of time we use _____.

energy

power

[43]

An electric iron rated at 1000 watts and operated for one hour would use $1000 \times 1 = 1000$ watt-hours of energy. How much energy would be required if this iron were used for two hours?

2000 watt-hours

Explanation: $1000 \times 2 = 2000$ watt-hours.

[44]

Just because we call the unit of energy a watt-hour, it is not necessary that power be used for a full hour before it can be considered energy. See the example below:

1000-watt iron used for 2 hours = 2000 watt-hours;

1000-watt iron used for 1/2 hour = 500 watt-hours;

600-watt bulb used for 1 minute or 1/60 hour = _____ watt-hours.

10

[81]

Even in a large wire the electrons will be bumping each other and the side walls of the wire to some extent. Therefore, if we increase the length of this wire, we also increase this bumping around of the electrons, which will eventually hinder their free movement. If we increase the _____ of the conductor, we will be adding some resistance to the free flow of the free electrons.

[81]

length

[82]

But regardless of what material we use as a conductor, this material will have a certain amount of resistance since the _____ will bump each other and rub the conductor wall to some degree even in the best conductors.

[82]

electrons

[83]

The size, length, and diameter of a conductor and the material of which it is made determine its number of f_____ _____ and its total _____.

[83]

free electrons

resistance

[84]

The amount of electric current that can flow through a conductor depends on the voltage supplied to the conductor and the size or _____ of the conductor.

[84]

resistance

[85]

A material with a very high resistance to electric-current flow is called an insulator. A conductor is the opposite of an _____.

[85]

insulator

[35]

240

Notice that in Panel VIII-C 120 volts are obtained on the secondary by using only 1/2 of the main secondary winding, whereas _____ volts are obtained from the total winding.

---

[36]

meter

We should now mention our electric meter and its function. The entire purpose of our meters is to measure the amount of energy that a customer uses to operate his appliances. The amount of energy that is used determines the amount of our customer's bill and is measured by his _____.

---

[37]

time

Earlier in this text we discussed power and made the statement that this is what makes your electric meter go around and determines your bill. To be exactly correct, we should say that our bill is determined by the length of time we use this power. Therefore, energy equals power times the _____ the power is used.

---

[38]

energy

If we lift a ten-pound weight, we need power. If we continue to lift this weight up and down, we use energy and get tired. If we lift this weight for 60 minutes, we get more tired than if we lift it for only one minute. Therefore, the longer that we use power, the more _____ is required.

---

[39]

energy

Electric energy may be considered in the same way. When we turn on a bulb, motor, etc., we require electric energy. If we use this device for 60 minutes, we require more _____ than we do if we use it for one minute.

[86]

For example, the cord on a household iron has a covering to prevent current from flowing from the wire to the person using the iron. This covering is the _____ on the cord.

[86]

insulator / insulation

---

[87]

A _____ will allow current to flow, while an _____ will prevent current from flowing.

[87]

conductor

insulator

---

[88]

At this point we need to briefly discuss the concept of direction of current flow. (Refer to Panel I-L.) When we connect the terminals of a battery by a conductor, the electrons from the [− /+] terminal will flow through the conductor to get to the [−/+] terminal.

[88]

−

+

---

[89]

(Refer to Panel I-L.) We see, therefore, that electrons will flow from − to _____ through a _____ .
     (Sign)

[89]

+

conductor

---

[90]

Due to the misunderstanding that occurred when electricity was first studied, current was believed to flow from + to − in a conductor. This direction would be [the same as/opposite to] that of electrons through a conductor.

[90]

opposite to

[31]

Thus, it is easy to recognize the type of electrical service provided a customer. A service drop consisting of only _____ wires will be carrying only 120 volts, while a three-wire service drop will be providing _____/_____-volt service.

[31]

two

120/240

[32]

Because we need to have 120/240 volts for some of our appliance needs in our home, the service drop will contain _____ wires.

[32]

three

[33]

Due to the way that transformers are made and connected, we are able to obtain this dual voltage of 120 or 240 volts from the secondary winding. Panel VIII-C shows how the windings are actually connected and Panel VIII-D shows an exterior view of such a transformer. In both cases the primary voltage is _____ volts and the secondary voltages are _____ and _____.

[33]

4800

120

240

[34]

An over-all view of an actual service entrance is shown on Panel VIII-E. Refer to the panel and describe the items which correspond to the letters below. (Items I and J have not yet been discussed, but try and guess what they might be.)

A ——————————— F
B ——————————— G
C ——————————— H
D ——————————— I
E ——————————— J

[34]

A. Transformer
B. Secondary circuit
C. Service drop
D. Transformer primary
E. Transformer secondary
F. Primary circuit
G. Meter
H. Fuse
I. Lightning arrester
J. Weather head

[91]

Therefore, although current is the movement of electrons, <u>by definition</u> current is shown to flow in the direction opposite to electron flow in a conductor. If this seems confusing, don't worry about it. Just remember that

    (1) current is the movement of electrons;
    (2) current is defined as flowing in a direction
            which is opposite that of an electron;
    (3) current flow in a circuit is from + to −.

[91]

[92]

In the following circuit indicate the direction of electron and current flow.

[92]

[93]

In an electric circuit the current flows from ____ to ____.
                                 (Sign)   (Sign)

Show this by a simple circuit diagram below.

[93]

+
−

[94]

From this point on we will always deal with <u>current flow</u> rather than electron flow. Therefore, it will be shown flowing from ____ to ____ in our circuits.
(Sign) (Sign)

[94]

+
−

[26]

For economical reasons companies try to serve more than one customer from a typical pole transformer installation. This can be accomplished by connecting wires to the secondary winding of one transformer and running these wires to several customers. (See Panel VIII-B.)

---

[27] secondary

The wires previously mentioned which are connected to the second-ary winding of the transformer are referred to as _____ distribution lines. (See Panel VIII-B.)

---

[28] service-drop

Then the wires connecting the customer's home and the secondary wires are referred to as service drops. The voltage enters the customer's home through the _____ - _____ wires.

---

[29] 120
240

As seen on Panel VIII-B, there are three wires coming from the secondary line to the house. These three wires carry our normal house voltage of 120/240 volts. The service drop to the house has a voltage of ____ and ____ .

---

[30] three

Most of the services to residential customers on the utility's system now are three-wire, 120/240-volt services. However, there are still a few service drops composed of only two wires. These provide only 120-volt service. Therefore, any home which needs and uses 120/240-volt electrical service must have a ser-vice drop of _____ wires.

II Ohm's Law

[1] Earlier in this course you learned that all <u>conductors</u> have some resistance although some have less than others. You also found that a common wire is a form of a _____.	**[1]** conductor
[2] Since all conductors have some resistance and since a wire is a conductor, all wires have some _____.	**[2]** resistance
[3] All electrical devices such as bulbs, motors, toasters, etc., have some form of wires in them. Since wires have resistance, electrical devices must have _____.	**[3]** resistance
[4] An electric <u>circuit</u> results when a source of voltage (electric potential) is connected to an electrical device by means of conductors. Therefore, before we can have a flow of electrons or current, we must have an electric _____.	**[4]** circuit
[5] All electrical devices have some resistance, which means that if these devices are connected by conductors to a source of voltage (electric potential) so that a _____ flows, we have an electric circuit.	**[5]** current

[21] To be absolutely correct, we should always use the equation $P = I \times E \times$ <u>power factor</u>. We will not attempt in this course to develop this concept any further. It is very important, however, that any time you attempt to make power or current calculations for electrical devices you first determine the _____ _____ of the device.	[21] power factor
[22] The formula $P =$ ____ gives us the formula $P =$ ____ $\times$ _____ _____	[22] IE IE $\times$ power factor
[23] Each individual electrical device will have its own _____-_____ rating.	[23] power–factor
[24] Typical relative sizes of system transformers may be seen in Panel VIII-A. Typical outputs for distribution pole-type trans-formers, like those used on your street, would be 5, 10, 25 or maybe 50 kva. A 10-kva transformer rated at 250-volt secondary could supply how much current before becoming overloaded? I (max) = _____	[24] 40 amp
[25] Since transformers contain coils of wire, if they become over-loaded, the wires may get _____ and become damaged.	[25] hot

[6]

When you pull the light switch in your car, your headlights come on. The lights are electrical devices and, therefore, have resistance; they are connected by wires to your battery, which is a source of electrical potential. Your battery, lights, and car wiring, when combined, make an _____ _____.

[6]

electric circuit

[7]

A light bulb, being an electrical device, will contain wires which have _____. When connected by conductors to a _____ source, this bulb forms part of an electrical circuit.

[7]

resistance

voltage

[8]

To have an electric circuit in which current can flow, we must have three things: source voltage, r_____ce, and c_____s.

[8]

resistance

conductors

[9]

A battery is an example of a source of electric potential, which we will henceforth refer to as <u>voltage</u>. Your car battery, therefore, is a source of electric potential or _____ for the lighting circuit of the car.

[9]

voltage

[10]

To have an electric circuit we must have a _____ connected by _____(s) to a _____ _____. (Refer to Panel II-A.)

[10]

resistance / load

conductor(s) / wires

voltage source / electric potential

[16]

Transformers are rated in size by the term kva. The term "k" is merely a symbol meaning 1000; therefore, 1 kva equals _____ va.

1000

[17]

The term "va" means volt × amp. Therefore, if a transformer supplies 200 amp at 250 volts, it must be _____ va in size. This would be the same as _____ kva.

50,000

50

[18]

The term volt-amp may seem to be the same as the term watt, which we discussed earlier. In many cases, watts and volt-amp may be the same. To be technically correct we should say that

$$\text{watts} = \overline{\text{volt-amperes}} \times \overline{\text{power factor}}$$

[19]

By the same token we could say

$$\text{volt-amperes} = \frac{\text{watts}}{\text{power factor}}$$

This power factor is a number which will always be between 0 and 1, that is, 0.5, 0.4, etc. Every individual electrical device has a _____ - _____ rating.

power-factor

[20]

Any time we attempt to determine the exact amount of current an electrical device will use or how much power it will require, we can get a general idea from the equation $P = I \times$ _____.

E

[11]

You will recall that in previous discussion conductors were described as materials through which current flows easily. While metal is the most common type of conductor, there are others, such as earth, impure water, and carbon, which will allow various amounts of current flow. [Note: This is not pure water which has been distilled or demineralized.] (See Panel II-B for examples of conductor materials.)

[11]

---

[12]

You can think of the metal in the body of your car as a conductor. The metal case of your flashlight is also a _____.

[12]

conductor

---

[13]

We saw in previous frames that a good conductor contains a very small resistance; a poor conductor contains a large amount of resistance. A piece of wood, which has a very large resistance, is a _____ conductor.

[13]

poor / non-

---

[14]

Unless otherwise specified, the remainder of this text will be concerned only with good conductors. They will, therefore, have _____ _____ resistance.
        (Own words)

[14]

very little / small

---

[15]

In fact, we will go one step further and consider our conductors to have zero (no) resistance. This is technically not possible since all conductors have some _____. For the time being, however, this assumption will make our work easier.

[15]

resistance

[11] All conductors have power loss due to their inherent resistance, since $P = \underline{\hspace{2cm}} \times R$.

[11] I^2

[12] Therefore, by increasing the transmission voltage we can decrease the transmission current required and decrease the $\underline{\hspace{2cm}}$ of the transmission lines.

[12] power loss / voltage drop

[13] A general rule to follow is that the longer the line over which power is transmitted, the higher the voltage should be. Some transmission lines which run many miles use 138,000 and 46,000 volts, whereas distribution circuits strung along streets may use only 4800 volts. Our house circuits, which may be only 100 feet long, use only 120 volts. (See Panel VIII-A.)

[13]

[14] Although very high voltages are good for transmission purposes, they are not practical for distribution circuits running along our streets, for there are increased problems with these high voltages and we don't need them for the shorter distribution circuits. We therefore use a $\underline{\hspace{2cm}}$ - transformer to reduce these high transmission voltages to the lower distribution voltages.

[14] step-down

[15] The larger step-up or step-down transformers are located in areas called substations. One purpose of a substation, therefore, might be to $\underline{\hspace{2cm}}$ voltage from transmission to distribution levels. (See Panel VIII-A.)

[15] step down

[16]

In electrical circuits, various symbols are used to represent the components. These are shown in Panel II-C. Refer to this panel as necessary for the remainder of this course.

[16]

---

[17]

An electrical circuit could be represented pictorially as follows.

Voltage source — Conductors — Resistance

Note that two conductors are required in a circuit. As will be seen later, this is due to the fact that current must be able to flow <u>from</u> and return <u>to</u> the voltage source. Represent this circuit symbolically, using the symbols that are on Panel II-C.

[17]

or

---

[18]

To have a complete or closed circuit, we must have an unbroken conductor coming from the voltage source to the electrical device and then returning to the source from the device. The circuit we drew in Frame 17 is an example of a _____ circuit.

[18]

closed

---

[19]

Refer to Frame 17. If we were to break one of the conductors connecting the source voltage to the device, we would have an open circuit, which is the opposite of the _____ circuit we had previously.

[19]

closed

---

[20]

The following circuit is an example of an _____ circuit. (Refer to Panel II-C for symbols, if necessary.)

[20]

open

[6]

In each cycle made by the conductor through the magnetic field, there are _____(Number)_ spots where no voltage is induced.

[6] two

[7]

Therefore, during the generator's cycle, the voltage is alternately turned on and off very rapidly. Hence, we get the name _____ for this kind of voltage.

[7] alternating

[8]

Most utility company generators develop electricity at a frequency of 60 cycles per second and, therefore, we call this electricity _____-cycle ac.

[8] 60

[9]

Because some generating plants are located long distances from major cities, it is economical to use high voltages on transmission lines. Some transmission voltages today range as high as 500,000 volts. The modern generators, however, generate voltage at only about 15,000 volts. To raise the level of voltage from the generation to the transmission level, we must use a _____ transformer. (See Panel VIII-A.)

[9] step-up

[10]

The higher the voltage used, the less current necessary to transmit the same amount of that power. You can see this for yourself from the following example. To transmit 14,000,000 watts of power (actually this is a small amount of power), how much current would be required if we transmitted with:

(1) 500,000 volts? I = _____

(2) 14,000 volts? I = _____

[10]
(1) 28 amp

(2) 1000 amp

[21]

A switch may be considered as two metal strips or conductors which can be made to touch or separate by being switched "on" or "off." When the switch is "on," the metal strips touch and current flows through the switch; when the switch is "off," the contacts (metal strips) are separated and current [can/cannot] flow.

[21]

cannot

[22]

A <u>switch</u> is used to "open" or "close" a circuit. Therefore, circuit (a) is _____ and circuit (b) is _____.

(a)   (b)

[22]

closed

open

[23]

Draw a circuit with all required components, showing a switch in the "off" position. Refer to Panel II-C, if necessary. Will current flow? Why? _____

[23]

No. Open circuit.

[24]

You have seen that an electric circuit consists of a voltage and an electrical device having resistance, both of which are connected by _____.

[24]

conductors

[25]

When the conditions mentioned in the previous frame are met, we will then have _____ flowing in the circuit.

[25]

current

VIII Summary

[1]

You have now completed the study of the basic concepts of the use of electricity. On the back cover is a reference card intended to summarize the more important concepts we have discussed. Tear it out and refer to it as necessary for this discussion of some of the more common practical uses of electricity and as reference for the terms we will use in this discussion.

[2]

Refer to Summary Panel VIII-S (1-2). The remainder of this course will deal with practical applications of this knowledge in relation to our type of work. Refer to any panels, previous frames, or summary cards necessary to help you answer the following questions. (Refer to the index to locate previous frames.)

[3]

All of the electricity used in utility company systems comes from generating plants over transmission lines to distribution lines and then to our homes. Panel VIII-A may help you understand this system design.

[4]

A generator makes voltage by mechanically causing conductors to move through _____ _____ .

> magnetic fields / lines of force

[5]

The generating station supplies _____ to our system. In this generating plant, machines generate _____ mechanically.

> voltage / power / electricity
>
> voltage / power / electricity

134

[26]

We have stated that current must be able to flow from and return to the voltage source, which requires two conductors. We have also stated that substances other than wires may be used for a

_____ .

[26]

conductor

---

[27]

Earth may be used for one conductor of a circuit. It must be true, therefore, that earth is a _____ of current.

[27]

conductor

---

[28]

In the following picture, we have driven rods into the ground or earth as noted. Will current flow? Why? (Refer to previous frame, if necessary.)

[28]

Yes. Ground is conductor and switch closed.

---

[29]

We have therefore used the _____ as a different type of conductor.

[29]

earth / ground

---

[30]

To make it easier to draw, we represent the connection to ground from the voltage source and resistance as in the diagram below.

Would current flow in this circuit? Why?_____

[30]

Yes. Ground is conductor and switch closed.

[86]

If the primary winding has 100 turns and the secondary winding has 50 turns, then the _____ is 2/1 and the secondary voltage will be 1/2 of the _____ voltage.

[86]

ratio

primary

[87]

If the secondary winding of a transformer has 4 times as many turns as the primary winding, this would be a step-_____ trans-former, and the secondary voltage produced would be _____ times as large as the primary voltage.

[87]

up

4

[31]

Draw an open and a closed circuit and label them (a) and (b), respectively. Show a voltage source, resistance, and switch, and use ground as one conductor. In which circuit will current flow? Explain.

[31]

(b) Because the switch is closed there is current. In (a) there is no current because of open switch.

[32]

Now draw a circuit in which current will flow, using two wires as conductors.

[32]

[33]

We have noted that although a circuit requires two conductors to be complete, these conductors need not always be wires. We have used _____ as a different type of conductor.

[33]

ground / earth

[34]

In our previous discussion we have seen that as the voltage of the circuit increases, the amount of current flowing in that circuit will also increase. Therefore, the amount of current flowing in a circuit depends on the _____.

[34]

voltage

[35]

You will recall that the resistance of the wire in the electrical device directly affects the current flowing through the device. If we increase the resistance of the circuit, the current flowing in the circuit will decrease. The current flowing in the circuit, therefore, is related to the _____ of the circuit.

[35]

resistance

[81]	**[81]**
If we wanted to step down 4800 volts to 240 volts, then the secondary winding would have fewer turns than the _____ winding.	primary
[82]	**[82]**
If the primary winding of a transformer has 100 turns and the secondary winding has 50 turns, the secondary voltage will be [greater than/less than/the same as] the primary voltage.	less than
[83]	**[83]**
If the primary has 100 turns and the secondary 1000 turns, then the secondary voltage will be [greater than/ less than/ the same as] the primary voltage.	greater than
[84]	**[84]**
The amount by which the voltage is stepped down or up is dependent on a mathematical <u>ratio</u> between the primary winding turns and the secondary winding turns. It is not important in this text that we work with this <u>ratio</u>, so long as we know there is a _____ between the turns in the windings.	ratio
[85]	**[85]**
The amount by which the voltage is stepped up or stepped down depends on the mathematical _____ between the number of turns on the primary and secondary windings of the transformer.	ratio

[36] To determine the amount of current in a circuit, therefore, we must consider the _____ and _____ of the circuit.	**[36]** voltage resistance
[37] As the applied voltage increases in a circuit, the current flow will _____ .	**[37]** increase
[38] As the resistance of a circuit increases, the current flowing in in the circuit will _____ .	**[38]** decrease
[39] Try to answer the following questions. (1) The unit of measure for current is _____ . Current is designated by the letter ___ . (2) The unit of measure for resistance is _____ . Resistance is designated by the letter ___ . (3) The unit of measure for voltage is _____ . Voltage is designated by the letter ___ .	**[39]** (1) amperes / amp I (2) ohms R (3) volts E or V
[40] Both E and V were used to represent voltage. In this text, E will represent source voltage, often referred to as electromotive force or emf. Some references will also use a script E, ε, to represent emf. The symbol V will be explained later as representing voltage drop across resistances. Both represent "volts" and should be treated the same in Ohm's law. If you completed the preceding frame correctly, <u>skip</u> to Frame 45. If you missed any of the answers, continue with Frame 41.	**[40]**
[41] Before proceeding, we should review the common terms and units for the parts of an electric circuit. (1) The amount of resistance in a circuit is measured in ohms. (2) The amount of potential difference in a circuit is measured in <u>volts</u>. (3) The amount of current in a circuit is measured in <u>amperes</u>.	**[41]**

VII - MAGNETISM 131

[76]

[76] voltage winding

Therefore, when we connect a transformer to the lines of our electric systems, the alternating voltage from our mechanical generators causes a constantly moving or fluctuating magnetic field from the primary winding. This causes a constant induction of _____ in the secondary _____ of the transformer.

[77]

[77] secondary

Any time we have an alternating voltage on the primary of a transformer, we will have a voltage induced in the _____ winding.

[78]

[78] more

We now have the method for inducing voltage from one winding to another winding. But how do we step down or step up the induced voltage? To step down the voltage we must have fewer turns in secondary winding than the primary winding. To step up the voltage we must have _____ turns in the secondary winding.

[79]

[79] fewer

To step down voltage by use of a transformer, there will be fewer turns in the secondary winding than in the primary winding.

[80]

[80] more

To step up voltage by use of a transformer, there will be _____ turns in the secondary than in the primary winding.

[42]

Now label the circuit below with the common units of measure.

(1) Flow of current _____

←(2) Resistance _____

(3) Voltage source _____

[42]

(1) amperes

(2) ohms

(3) volts

---

[43]

We have applied the common units of amperes, volts, and ohms to the terms current, potential difference, and resistance, respectively, of the electric circuit components. We should now assign the letters or symbols to be used for these terms. They are merely convenient designations for parts of a circuit, not initials for any words. Fill in the common units below.

    (1) Current   _____ (letter is I)

    (2) Potential difference _____ (letter is E)

    (3) Resistance   _____ (letter is R)

[43]

(1) amperes

(2) volts

(3) ohms

---

[44]

To summarize our use of common terms and letters, complete the following.

	Unit	Letter
(1) Current	_____	____
(2) Potential difference	_____	____
(3) Resistance	_____	____

[44]

(1) amperes   I

(2) volts   E

(3) ohms   R

---

[45]

We will designate the flow of current in a circuit by the use of arrows to show which way the current is flowing. Refer to the circuit below.

The current is flowing in a [clockwise/counterclockwise] direction.

[45]

clockwise

[71]

We require a source voltage to force current through the primary windings. This source voltage originates at a generator similar to the one described in Frames 35-54. You will recall that during each cycle, or 120 times per second, the generator was in a "dead" position, or let's say that no voltage was in _____ the conductors.

[71] induced

---

[72]

If our source voltage was off 120 times per second, then our primary voltage will be off _____ times per second.

[72] 120

---

[73]

You will also recall that as we turn the current on and off in the primary winding, this will cause a movement of the magnetic field past the secondary winding. As the magnetic field moves past the secondary winding, a _____ will be induced in this winding.

[73] voltage

---

[74]

Therefore, in order to induce a voltage in the secondary windings of a transformer, we need only cause a fluctuating or alternating (off-on) magnetic field at the primary winding. This can be done by causing a fluctuating voltage across that _____ winding.

[74] primary

---

[75]

This fluctuation, or alternating primary voltage input, automatically exists due to our method of mechanical generation of power. Remember that a mechanical generator has an alternating output voltage since the voltage is off two times a cycle or _____ times per second.

[75] 120

[46]

Draw the following circuit: closed circuit with a battery, resistance, switch, and clockwise current. Label with symbols. Refer to Frame 45, if necessary.

[46]

[47]

State the mathematical relationship between voltage, current, and resistance and solve the following circuit for current flow.
I = ___ amp.

[47]

$$I = \frac{E}{R}, \quad E = IR, \text{ or } R = \frac{E}{I}$$

$I = 2$ amp

[48]

If you answered the preceding frame correctly, proceed to Frame 68. If you missed or do not understand any part of it, proceed with the next frame.

[48]

[49]

We can therefore summarize that the current which will flow in any closed circuit is related to the r_____ and the source v_____ of the circuit.

[49]

resistance

voltage

[50]

To review, we found that by either increasing the source voltage or decreasing the circuit resistance, we caused the circuit current to increase. Therefore, there is a definite relationship, called Ohm's law, between voltage, _____, and _____ in a circuit.

[50]

current

resistance

(in either order)

[66] If 4800 volts were to be decreased to 240 volts, then a _____ transformer would be needed, and the 4800 volts would be connected to the _____ winding.	[66] step-down primary
[67] If we stepped down 4800 volts to 240 volts, then the 240 volts would come from the _____ windings.	[67] secondary
[68] The fact that electric current flowing in a conductor causes a magnetic field is also considered in transformers. A magnetic field will _____ a voltage in a conductor when the lines of force of the magnetic field move past the conductor.	[68] induce
[69] As current flows through the primary winding of a transformer, there will be a _____ established. (See Panel VII-I.)	[69] magnetic field
[70] Now, if we place the secondary winding of the transformer near enough to the primary winding, it will intersect the magnetic field from the primary. By making this primary magnetic field turn off and on, it will appear to be moving in relationship to the secondary winding. This moving magnetic field would induce a voltage in the _____ winding. (See Panel VII-J.)	[70] secondary

II – OHM'S LAW   30

The current in a circuit always equals the value obtained when
the source voltage is divided by the circuit resistance. Therefore, if we have a source voltage of 10 volts and a circuit resistance of 2 ohms, we divide the voltage by the resistance and
determine the current flow to be _____ amp.
(Number)

[51]

5

---

[52]

The expression for Ohm's law says that circuit current equals
source voltage divided by circuit resistance. Now write this
relationship, using the <u>common units</u> for circuit components.

[52]

$$\text{amperes} = \frac{\text{volts}}{\text{ohms}}$$

---

[53]

Recalling that the letters I, E, and R stand for current, source
voltage, and resistance, respectively, we can write the Ohm's-
law expression in a more formal arithmetical form. (Refer to
previous frame, if necessary.)

___ = ___

[53]

$$I = \frac{E}{R}$$

---

[54]

Since we have seen that I = E/R (current equals voltage divided
by resistance), we can always determine the amount of current
flowing in a circuit if we know the voltage and _____.

[54]

resistance / ohms

---

[55]

If we have a source voltage of 10 volts and a circuit resistance of
2 ohms, we can determine our circuit _____. In addition determine the values for the following.

E = _____, R = _____.

[55]

current

10 volts

2 ohms

[61]

primary

[61]
The voltage that we want to change (that is, increase or decrease) is always connected to the primary winding. If we want to decrease 4800 volts to a lower voltage, we would connect it to the _____ winding of a transformer.

---

[62]

secondary

[62]
The new voltage we get from the transformer (that is, the increased or decreased voltage) comes from the secondary winding. If we decreased 4800 volts to 240 volts, then the 240 volts would come from the _____ winding.

---

[63]

step-down

[63]
Instead of saying "decrease of voltage" when speaking of transformers, we say, "step-down voltage." If we have a transformer that decreases voltage, then we have a _____ - _____ transformer.

---

[64]

secondary

step-down

[64]
To change a primary voltage of 4800 volts to a lower or voltage of 480 volts, we would need a _____ - _____ transformer.

---

[65]

step-up

[65]
To increase voltage means to step up voltage when speaking of a transformer. To increase 240 volts to 4800 volts, we would need a _____ - _____ transformer.

[56]

Using the values in Frame 55, determine the amount of current in the circuit mentioned. Remember that I = E/R. Therefore,

I = _____ amp.

[56]

$I = \dfrac{10}{2} = 5$ amp

---

[57]

Solve the following. Given that R = 10, E = 100, find I.

I = ___ amp

[57]

10

---

[58]

Write the relationship between current, voltage, and resistance in a circuit.

$$\frac{\quad}{\quad} = \frac{}{R}$$

[58]

$I = \dfrac{E}{R}$

---

[59]

Write the same relationship between current, voltage, and resistance in these terms.

Current = ———————

[59]

$\dfrac{\text{voltage}}{\text{resistance}}$

---

[60]

Write the same relationship as in Frames 58 and 59, using the common units of amperes, ohms, and volts for the circuit components.

$$\frac{\quad}{\quad} = \frac{\text{volts}}{}$$

[60]

amperes $= \dfrac{\text{volts}}{\text{ohms}}$

[56] decrease	**[56]** The purpose of a transformer is to change the voltage which exists on our conductors (Panel VII-F) to the value we desire. Transformers then must be able to increase or _____ the value of voltage available.
[57] transformer	**[57]** The wire conductors at the top of the poles in your neighborhood have a voltage level of about 5000 volts, which is much too high for our home use since we need only 120 or 240 volts. Therefore, to decrease the voltage we would need a t_____ .
[58] windings coils	**[58]** A transformer consists of two sets of windings (a coiled conductor) placed very close together. A transformer has two sets of _____ or _____ of wire. (See Panel VII-F and -G.)
[59] voltage	**[59]** Transformers come in many sizes, shapes, and types, some of which may be seen in Panel VII-H. Regardless of what they look like, however, they all contain windings of wire and are used to change the level of _____ .
[60] two primary secondary	**[60]** These windings are called the primary and secondary windings. A transformer, therefore, has _____ kinds of windings referred to as _____ and _____ .

[61]

Now write the relationship between current, voltage, and resistance in three different forms.

[61]

$$\text{current} = \frac{\text{voltage}}{\text{resistance}}$$

Current equals voltage divided by resistance.

$$\text{amperes} = \frac{\text{volts}}{\text{ohms}} \; ; \; I = \frac{E}{R}$$

---

[62]

From this point on, when we use the Ohm's-law relationship, we will use the letters which represent the circuit components. Write this relationship and use it to solve for current in the circuit below.

I = _____

[62]

2 amp

$$I = \frac{E}{R} = \frac{4}{2} = 2 \text{ amp}$$

---

[63]

This relationship you have learned, Ohm's law, will be true for all electric circuits. Solve the following circuits for the amount of current.

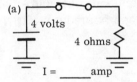 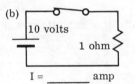 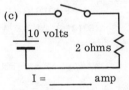

[63]

(a) 1 amp, (b) 10 amp, (c) 0 amp

(If you missed (c), remember that current cannot flow in an open circuit. If the switch had been closed, the current would have been 5 amp.)

---

[64]

The relationship of current, voltage, and resistance can be expressed in different ways, all of which will be true for any electric circuit. One of these that you learned is _____.

[64]

$$I = \frac{E}{R}$$

---

[65]

Two other ways of writing this same relationship of current, voltage, and resistance (which we will refer to frequently as an equation) are

$$E = IR \quad \text{and} \quad R = \frac{E}{I} .$$

Now write the third way of stating the equation.

[65]

$$I = \frac{E}{R}$$

[51]

We have been referring to an on-off voltage, which we could call alternating voltage, such as that produced by a mechanical gener-ator. This alternating voltage would cause an alternating current to flow in a circuit since current depends to a large extent on

_____ .

[51]

voltage

[52]

Because this alternating voltage causes an alternating current to flow, we call it an alternating-current voltage. The abbreviation for alternating-current voltage is ac-voltage. If we have ac-voltage, it means that our voltage is _____ing.

[52]

fluctuating / alternating

[53]

Utility companies use this alternating voltage. We saw earlier that since the mechanical generators make 60 revolutions or cycles per second, we have 60-cycle electricity. Because this voltage is alternating, we would correctly call this voltage 60-_____-voltage.

[53]

cycle ac

[54]

By ac-voltage we mean _____ing voltage, and on the lines this voltage is off 120 times per _____ .

[54]

alternating

second

[55]

Refer to Panel VII-F. We have studied the method of producing electricity at the generating plant and how the wire conductors on the utility pole carry this electricity. On the panel there is another device that this electricity must go through before the house is served with electrical power. This item is the trans-former, which is our next topic.

[55]

[66]

To make the solution for any one of the letter symbols as simple as possible, we have these three forms of Ohm's law, in which the symbol for which we are solving is always on the left of the equal-sign. These forms are listed on Panel II-D for your convenience. From now on refer to this panel whenever you need to use Ohm's law in any form.

[66]

---

[67]

Complete the following.

$$I = \frac{E}{R}, \quad \underline{\quad} = IR, \quad \underline{\quad} = \frac{E}{I}.$$

[67]

R

E

R

---

[68]

Rewrite the following forms of the equation, using the common units for the letters shown.

$$I = \frac{E}{R}, \quad E = IR, \quad R = \frac{E}{I}.$$

[68]

$$amperes = \frac{volts}{ohms}$$

$$volts = amperes \times ohms$$

$$ohms = \frac{volts}{amperes}$$

---

[69]

Now write all three forms of the equation.

$$I = \underline{\quad}, \quad R = \underline{\quad}, \quad E = \underline{\quad}.$$

[69]

$$\frac{E}{R}$$

$$\frac{E}{I}$$

IR

---

[70]

As you have observed, there are three components which will be found in any complete electric circuit: current (amperes), difference of potential (volts), and resistance (ohms). Therefore, if we know two of these values, we can determine the third by arithmetic. Here is an example: If we have a circuit with 12 volts and 3 ohms, what is the current (amperes)? _____

[70]

4 amp

Explanation: $I = \frac{E}{R}$;

therefore $I = \frac{12}{3} = 4.$

[46] induced	[46] You should have observed from the pictures in VII-E that the time during which voltage is not being induced is very small compared with the time it is being induced. The conductor X cuts the magnetic field continuously as it proceeds from position A to position C to position E (A). When the conductor cuts through this magnetic field, voltage is _____ .
[47] voltage	[47] The only time voltage is not being induced is during the instant that conductor X is at the exact position A or C. When at this position, the conductor is not moving past the magnetic field and _____ is not being induced.
[48] on / being induced	[48] Therefore, although the voltage is not being induced or is "off" 120 times per cycle, it is _____ for a larger period of time than it is off.
[49] off on on off	[49] In one cycle our voltage would appear in the following sequence: on, [on/off], [on/off] off. It would be [on/off] for a longer period of time than it would be [on/off], however.
[50] on	[50] Note that although the voltage is turning off 120 times per second, the voltage is on for so much longer a period of time than it is off that it appears to us to be constantly _____ .

[71]

When the parts of the circuit have the following values, see if you can find I for the three cases.

     (1) 4 ohms, 20 volts; I = _____ amp

     (2) 20 ohms, 4 volts; I = _____ amp

     (3) 3 volts, 3 ohms; I = _____ amp

(Note that the current may be less than one ampere.)

[72]

When it is necessary to find:

     (1) the voltage in a circuit, solve for the letter ___; in the formula _____ .

     (2) the current in a circuit, solve for the letter ___; in the formula _____ .

     (3) the resistance in a circuit, solve for the letter ___; in the formula _____ .

[73]

Work the following problems and show the form of the equation that you use.

     (1) A circuit has 4 amp and 8 volts. Find the ohms.

     (2) A circuit has R = 2 and I = 8. Find E.

     (3) A circuit has a voltage of 2 and a resistance of 2. Find the current.

[74]

Suppose that in an electric circuit there is a current of 4 amp. What will be the value of the current in the circuit when the original voltage to this circuit is doubled in value? _____

[75]

If we double the resistance of the circuit in which the current of 4 amp is flowing, the value of I then becomes ___ amp.

---

[71]

(1) $I = \frac{E}{R} = \frac{20}{4} = 5$

(2) $I = \frac{E}{R} = \frac{4}{20} = \frac{1}{5}$ (or 0.2)

(3) $I = \frac{E}{R} = \frac{3}{3} = 1$

[72]

(1) E; (E = IR)

(2) I; $(I = \frac{E}{R})$

(3) R; $(R = \frac{E}{I})$

[73]

(1) $R = \frac{E}{I} = \frac{8}{4} = 2$ ohms

(2) E = IR = 8 × 2 = 16 volts

(3) $I = \frac{E}{R} = \frac{2}{2} = 1$ amp

[74]

8 amp (4 doubled)

[75]

2

Explanation: $\frac{1}{2} \times 4$.

[41]

This mechanical device or generator is set up to rotate the conductors one complete revolution or _____ 60 times per second.

[41]
cycle

---

[42]

The mechanical generators of most power companies rotate their conductors through a cycle 60 times per second. Therefore, we say that the system has _____-cycle electricity.

[42]
60

---

[43]

Refer to Panel VII-E. During two positions in a cycle the conductors do not include a voltage. Which positions are these? [A/B/C/D]

[43]
A and C
(Note: A and E are the same position.)

---

[44]

You found two pictures on the panel in which no voltage was induced, namely A and C. (We hope that you did.) Therefore, if the conductors revolve at 60 cycles per second, and if at two times on each cycle we have "dead positions," then [100/110/120/140] times per second the generator is not producing voltage.

[44]
120

---

[45]

So far we have learned that a generator turns or rotates a conductor or conductors through a magnetic field, which induces a voltage in the conductor. This generator rotates the conductor 60 cycles a second, and 120 times a second the conductors are in "dead position," where no _____ is induced.

[45]
voltage

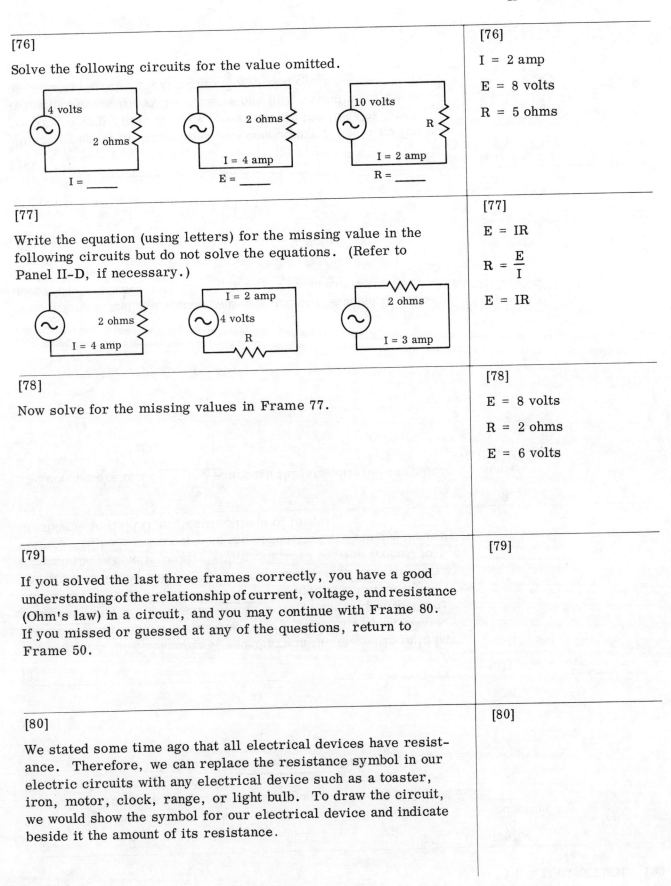

[76]

Solve the following circuits for the value omitted.

I = _____

E = _____

R = _____

[76]

I = 2 amp

E = 8 volts

R = 5 ohms

[77]

Write the equation (using letters) for the missing value in the following circuits but do not solve the equations. (Refer to Panel II-D, if necessary.)

[77]

E = IR

R = $\dfrac{E}{I}$

E = IR

[78]

Now solve for the missing values in Frame 77.

[78]

E = 8 volts

R = 2 ohms

E = 6 volts

[79]

If you solved the last three frames correctly, you have a good understanding of the relationship of current, voltage, and resistance (Ohm's law) in a circuit, and you may continue with Frame 80. If you missed or guessed at any of the questions, return to Frame 50.

[79]

[80]

We stated some time ago that all electrical devices have resistance. Therefore, we can replace the resistance symbol in our electric circuits with any electrical device such as a toaster, iron, motor, clock, range, or light bulb. To draw the circuit, we would show the symbol for our electrical device and indicate beside it the amount of its resistance.

[80]

[36]

This mechanical device is known as a generator. A mechanical force keeps the wire constantly moving past a magnetic field and this in turn keeps up a constant _____ of voltage in the wire. (See Panel VII-E for this frame through Frame 46.)

[36] induction

[37]

In position A the conductors X and Y are in the magnetic field but not moving through it. In moving from A through B to C, it was necessary that the conductors move through the magnetic field; therefore, voltage was induced in the conductors. In C, again, voltage will not be induced because the conductors are not moving through the magnetic field. With conductors moving from C to E, voltage will be induced because the conductors are moving through the magnetic field (i.e., cutting lines of force).

[38]

The conductors must _____ through the magnetic field to induce a voltage.

[38] move

[39]

In positions A and E a voltage was not induced because the conductors [were/were not] moving past the magnetic field.

[39] were not

[40]

Referring again to Panel VII-E, as conductors X and Y go from position A to position E, they move through the lines of force twice (positions B and D). This is one full revolution of the conductors or, in electrical terminology, one cycle.

[40]

[81]

Draw a circuit with a 4-volt generator, a 1-amp current, and a light bulb with the proper resistance to match.

R = _____ ohms.

[81]

(R = E/I = 4/1 = 4)

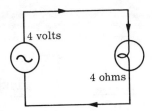

4 volts

4 ohms

[82]

Before going on, it is important that you associate the electrical terms with their proper unit and letter. Complete the following:

Term	Unit	Letter
(1) _____	volts	_____
(2) _____	_____	I
(3) resistance	_____	_____

[82]

(1) Potential difference E
(2) Current   amperes
(3) Ohms   R

(If you missed any of these answers, review Frames 41-44 before proceeding.)

[83]

Likewise, it is extremely important at this point to completely understand the Ohm's-law relationship. Therefore, using first the electrical units and then the letters, write Ohm's law.

(1) amperes = ——— 　　(2) ___ = $\frac{}{R}$

[83]

(1) amperes = $\frac{volts}{ohms}$

(2) $I = \frac{E}{R}$

(If you missed part of this question or feel you do not understand Ohm's law completely, review Frames 48-83 before proceeding.

[31]

We must have a moving magnetic field before we can induce volt-age in a conductor. We can obtain this moving field by moving either the conductor or the magnetic field itself, or by rapidly _____.

(Own words)

[31] turning the magnetic field on and off, thus causing it to change.

[32]

Now, if we set up two magnets with a magnetic field between the poles and pass a wire through the field, it will induce _____ in the wire.

[32] voltage / current

[33]

By passing a conductor through a _____ of two magnets so that the magnetic field is moving past the con-ductor, we are able to induce _____.

[33] magnetic field / voltage

[34]

If we constantly keep passing this wire through the magnetic field, there is a constant _____ of voltage.

[34] induction

[35]

Now all we need is a mechanical device to continually pass the conductor through the magnetic field, thus inducing a _____ voltage in the conductor.

[35] continuous

III Series Circuits

[1]

For convenience, let's refer to any electrical device we have in our circuit as a <u>load</u>. If we increase the number of light bulbs in a circuit, we are increasing the _____.

[1]

load / resistance

---

[2]

Since toasters, irons, motors, lights, etc., are all electrical devices containing wires, they have _____ and are referred to as electrical loads.

[2]

resistance

---

[3]

We generally say we are increasing our electrical loads as we [add/subtract] electrical devices to (from) our circuit.

[3]

add

---

[4]

There are two ways to add electrical devices or _____ to our circuit. They may be connected in what is commonly called "series" or "parallel." We will discuss the series method first.

[4]

load(s)

---

[5]

Definition: A <u>series</u> circuit is one in which the resistances or loads are connected end to end with the same current flowing through all the loads as shown below.

In this circuit there are five resistances or loads connected in _____.

[5]

series

[26]

So long as there is movement between a magnetic field and a conductor, regardless of which moves, there will be an _____ _____ in the conductor.

[26]

induced voltage / electric current

---

[27]

We can simulate movement of a magnetic field past a conductor by simply turning the field off and on rapidly. As our magnetic field is turned off and on, it appears to the conductor that the lines of force of the field are moving since they alternately appear and disappear. This apparent _____ of the lines of force induces a voltage in the conductor.

[27]

moving

---

[28]

When moving the lines of force in relationship to a conductor by turning them rapidly on and off, we can induce a _____ in an adjacent conductor.

[28]

voltage / force / current

---

[29]

Since current flowing in a conductor will cause a magnetic field, we can easily cause this magnetic field to move (turn on and off) by turning the _____ in the conductor on and off.

[29]

current

---

[30]

By alternately turning the conductor current on and off we can make a _____ing magnetic field which will _____ voltage in a conductor.

[30]

moving

induce

[6]

Some Christmas tree lights are connected as shown below.

Plug

These are connected in _____.

[6]

series

---

[7]

In the circuit shown in Panel III-A, you can see that although we have added five loads to the circuit, there is only one path for the current to take. It must flow through all the loads to return to the source. Since these loads are connected in _____, we can then say that in a _____ circuit the current flows in only one path. This will always be true.

[7]

series

series

---

[8]

Because the current in a series circuit can flow only in one path and all of our resistances or loads are connected end to end, it appears that in a _____ circuit the same current must flow through all of the loads in the circuit.

[8]

series

---

[9]

Since there is no way for the current to get out of the circuit, the amount of current that leaves the voltage source must also return to the voltage source. Since there is only one path for this current to flow in a series circuit, the same _____ must flow through all of the loads.

[9]

current

---

[10]

In a series circuit, therefore, the _____ are connected end to end and the same _____ will flow through all of the loads.

[10]

loads / resistances

current

[21]

Previously, we found that the magnetic field had to move by the wire conductor. If the magnet passes by the conductor, then the _____ _____ of the magnet is moving by the conductor.

[21]

magnetic field

[22]

Then it is not the magnetic field alone that causes the induction of voltage but the movement of the magnetic field past the conductor which produces the _____ _____ .

[22]

induced voltage / current flow

[23]

The magnetic field must _____ past the conductor to induce a voltage in a conductor.

[23]

move

[24]

Since it has been shown that the movement of a magnetic field past a wire causes an induced voltage in the wire, it follows that by moving the conductor through the magnetic field (see Panel VII-D) in a manner similar to that shown in Panel VII-C, we can again induce _____ in the conductor.

[24]

voltage

[25]

As a magnetic field moves past a conductor or a conductor moves past a magnetic field, we are able to _____ voltage within the conductor.

[25]

induce

[11] We noted before that resistance in a circuit helps to determine the amount of current flowing in the circuit. Think of resistance as in a garden hose. When you close the nozzle, you restrict the opening and therefore resist the flow of water. If you open the nozzle completely, there is very little _____ to the flow of water.	[11] resistance
[12] Adding resistance to a circuit is like closing the nozzle on the hose. If we keep the value of the voltage the same and add resistance in series to the circuit, the current decreases. This is true because the current must flow through all the resistance in the circuit. To see how much resistance the current must flow through in a series circuit, we simply add all the different load resistances to get a total circuit _____.	[12] resistance
[13] We can best see this from the circuits shown below. Circuit (a) has 2 ohms resistance. Circuit (b) has 2+2+2+2 or 8 ohms total resistance. Which circuit will restrict the flow of current more, (a) or (b) ___? Why?_____.	[13] (b) More resistance for current to flow through.
[14] The circuit resistance with which we are concerned when trying to determine current flow is the _____ resistance of the circuit.	[14] total
[15] Regardless of how many individual resistances are combined to make the total resistance, the amount of current flow in a series circuit, where the voltage is constant, will depend on the _____ resistance of the circuit.	[15] total

[16]
If we wish to force a current to flow in a conductor, we should move a conductor through the _____ of _____ of a magnetic field. (See Panel VII-B.)

[16] lines of force

[17]
Of course, we know that before current flows it has to have a force to push it and we are referring to this force as _____.

[17] voltage

[18]
Then a magnetic field actually sets up or induces _____ to push or force current flow.

[18] voltage

[19]
Suppose that we make a closed circuit of a wire and a voltmeter and pass a magnetic field close to the wire. The needle at the voltmeter would be deflected, indicating that voltage has been induced in the wire. (See Panel VII-C.) Therefore, when a magnetic field moves past a wire, there is _____ induced.

[19] voltage

[20]
Since, as seen in Panel VII-C, the magnet itself did not actually touch the conductor, it must be that the voltage was induced by the _____ of _____ of the magnet.

[20] lines of force

[16]

Find the total resistance in the circuits below and determine which one will restrict the current flow the most. Consider the voltage to be the same in all cases.

[16]

All 5 ohms.

All will restrict I the same.

[17]

Write Ohm's law and determine the current in the circuits below.

I = _____   I = _____   I = _____

[17]

(a) $I = \dfrac{E}{R} = \dfrac{10}{5} = 2$ amp

(b) $\dfrac{5}{20} = \dfrac{1}{4}$ amp

(c) 120 amp

[18]

All loads are connected end to end in a _____ circuit.

[18]

series

[19]

In a series circuit, the same current will flow through all _____ in the circuit.

[19]

loads / resistances

[20]

The total resistance in a series circuit equals _____ _____. (Own words)

[20]

the sum of all the individual resistances

[11]

See figure below. If current is being conducted through the wire, then the lines surrounding the wire are _____ of _____ or a _____ field.

Wire conductor

[11] lines of force magnetic

---

[12]

If we have an electric circuit, then we know that there is a _____ resulting from the electric current flow _____ in the circuit.

[12] magnetic field / force

---

[13]

We have found that the flow of electric current causes a magnetic field around its conductor. There appears, then, to be a definite relationship between magnetic fields and _____ _____ .

[13] electric current

---

[14]

We cannot have current flowing in a conductor without establish-ing a _____ _____ around the conductor.

[14] magnetic field

---

[15]

Since we cannot have a flow of electric current in a conductor without causing a magnetic field, then, we may expect, that by placing a wire within a magnetic field we can cause a _____ to flow in the wire.

[15] current

[21]

Recall that earlier we referred to voltage as a force that causes current to flow. When current flows through an electrical device or resistance, therefore, there must be a _____ to force the current to flow.

[21]

voltage

---

[22]

The current flows through the load due to the force or voltage across the load, which is the difference in potential between points A and B in Panel III-B. The voltage across a load can exist only between _____ points.

[22]

two

---

[23]

To repeat, voltage can exist only between two points. The difference of potential across a load that causes current flow is called the _____ _____ across the load. (Refer to Panel III-C.)

[23]

voltage drop / voltage

---

[24]

If we were to measure the voltage in a circuit, we would have to measure across a load or resistance, since voltage exists between _____ points separated by resistance.

[24]

two

---

[25]

Refer to Panel III-D. Voltage will appear between terminals A and B, A and C, and A and ___ but not between B and C or ___ and ___.

[25]

D

C and D   /   B and D

[6]	[6]
	Magnetism is an invisible force. Just as wind, for example, can provide a strong force though it is invisible, magnetism is an _____ force.
invisible	
[7]	[7]
	In watching a football game our interest is focused on the area of action. We know this area of action as the football field. The same holds true in studying the magnetic action of the magnet. Our interest is on the magnetic f_____.
field	
[8]	[8]
	The magnetic field about a magnet is best described as invisible lines of force leaving the magnet at one pole and entering it at another. (See Panel VII-A.) The magnetic _____ can be described as ____ ____ ____ ____.
field	
invisible lines of force	
[9]	[9]
	The points at which the lines of force leave or enter the magnet are the "poles" of the magnet. There are l_____ of _____ entering and leaving the n_____ and s_____ poles of the magnet.
lines of force	
north	
south	
[10]	[10]
	Years ago a Danish scientist, Oersted, found that if a compass was held near a wire conducting current, the compass needle was deflected. If the current was shut off, the needle returned to its original position. This scientist now knew that the electric current or electron flow was creating a magnetic field around the wire. This deflected the compass needle. From this little experiment, the scientist contended that electric current flow creates a _____.
magnetic field	

[26]

In order to have voltage in an electric circuit, there must be, in addition to the source voltage, a _____, across which the voltage can be measured.

[26]

load / resistance

---

[27]

You may understand this last statement better if you remember that voltage = current × resistance (E = IR) and, therefore, we cannot have voltage across a load unless we have both _____ and _____.

[27]

current

resistance

---

[28]

Since force is required to cause current to flow through a load, and since force and voltage are similar, _____ is required to cause current to flow through a load.

[28]

voltage

---

[29]

Again, since voltage and force are synonymous, our source voltage could also be considered a _____. Hence, source voltage is often called "electromotive force."

[29]

force

---

[30]

The only place from which our circuit can obtain voltage is from the source, since loads cannot make voltage. The voltage used in the circuit, then, can come only from the _____.

[30]

source

[1]

In previous frames we saw what electricity is, how it works for us in the form of power and the limitations we must observe in connection with its use, such as conductor size, resistance, high or low voltage, etc.

We will now study a new concept which is closely associated with the production and use of electricity. This is magnetism.

[2]

poles

In using natural magnets it was found that magnetism is concentrated at two points, usually the ends of the magnet. These points are called poles of the magnet. Magnetism is concentrated at the end points or _____ of the magnet.

[3]

poles

The earth is one big magnet, and its end points or the top and bottom of the earth are also known as _____.

[4]

north
south

The poles of the earth are known as north and south poles. Let's call the poles of any magnet _____ and _____ poles, the same as those of the earth.

[5]

north
south
poles

To sum up so far, magnetism is concentrated at the _____ and _____ of the magnet.

[31] A load cannot make voltage; it can only use it. Before current can flow through the loads, the loads need a voltage across them which will cause the current to flow through them. Before current can flow through a load, therefore, there must be a _____ across the load.	**[31]** voltage
[32] Since our source supplies all of the voltage available to the circuit, each load uses part of this voltage to cause _____ to flow through it.	**[32]** current
[33] Therefore, each load uses part of the _____ supplied from the source when current is caused to flow through the resistance of the load.	**[33]** voltage
[34] As the loads use part of the voltage from the source, we say that the source loses or <u>drops</u> part of its voltage at each load. It is normal, then, to call the voltage a load uses a voltage _____.	**[34]** drop
[35] If you look at Panel III-E, you will see that all of the voltage supplied to a circuit is used by the loads. Conversely, the loads cannot have more voltage than the amount _____ to the circuit from the source.	**[35]** supplied / available

[101]

When a 120-volt device is connected to a voltage of 240 volts, the wires of the device get very hot due to the _____ _____ of the device.

[101]

heat loss / power loss / high voltage

[102]

In fact, if a 120-volt device is connected to 240 volts, the wires will get so _____ that they will melt almost immediately.

[102]

hot

[103]

A high voltage condition will _____ the life of an electrical device and a very high voltage may cause the device to fail _____.

[103]

shorten

immediately

[104]

A motor is designed for operation at 120 volts. Explain fully the effect the following conditions will have on the motor.

    (1) 90 volts: _____

    (2) 115 volts: _____

    (3) 240 volts: _____

[104]

(1) Operate slowly or inefficiently.

(2) Operate normally.

(3) Have short life due to heating; probably burn out immediately.

[36]

We will now introduce a new symbol, V, which we will call voltage drop. We have seen that E represented s_____ v_____.
We will now see that V represents the amount of source voltage dropped at each _____ in an electric circuit.

[36]

source voltage

load / resistance

[37]

We will call this voltage dropped at each load the voltage drop of the load and will use for its symbol the letter ___.

[37]

V

[38]

Here we use V, the letter for voltage drop, exactly the same as we have used E, which stands for s_____ v_____.

[38]

source voltage

[39]

Since they are both voltages, they are treated alike. The letter V merely indicates that we are talking about the voltage across a _____, whereas E means we are talking about

_____ _____.

[39]

load / resistance

source voltage

[40]

A portion of the source voltage in a series circuit is dropped or used across each load. Therefore, the total voltage drops of all of the loads cannot exceed the _____ voltage.

[40]

source

[96]

Generally speaking, customers of a utility do not have high-voltage conditions unless they make a _____ in connecting their wires to the power company's wires.

[96]

mistake

---

[97]

What choice of voltages does the average customer have in his home? ____ or _____

[97]

120 or 240 volts

---

[98]

If the customer makes a mistake and connects a 120-volt device to the 240-volt electric wires, the life of the device will be

_____.

[98]

shortened

---

[99]

Since 240 volts would represent twice the correct voltage for a 120-volt device, this would represent an extremely _____ _____ condition and the device would fail almost immediately.

[99]

high voltage

---

[100]

Because heat loss of a device depends on the voltage (and resulting current) applied to it and since this heat loss occurs in the wires of a device, high voltage would cause the wires to get ____.

[100]

hot

[41]

The voltage drop across each load in a series circuit will depend on two quantities — the current through the load and the resistance of the load. We can, then, determine how much of the source voltage each load will use in a series circuit by the equation V = ___ × ___.

[41]

$I \times R$

---

[42]

Refer to Panel III-F for an example of voltage drop in a series circuit. In Part (a), the total resistance or R_T = _____.

[42]

30 ohms

Explanation:
$R_T = 10 + 20 = 30$ ohms.

---

[43]

Refer to Panel III-F, Part (a). Since R_T = 30 ohms, the circuit current will equal ___ amp.

[43]

2

Explanation:
$I = \dfrac{E}{R} = \dfrac{60}{30} = 2$ amp.

---

[44]

Refer to Panel III-F, Part (b). Since this is a series circuit, the same _____ will flow through all loads; it equals ___ amp.

[44]

current

2

---

[45]

Refer to Panel III-F, Part (b). The voltage drop of load A is found as follows: V = IR = 2 amp × 10 ohms = 20 volts.

Find the voltage drop of load B. _____

[45]

40 volts

[91]

The heat loss of a device will _____ as the current in it increases. Why? _____

[91]

increase

$P = I^2R$ and R is constant or fixed.

---

[92]

Therefore, if we subject a device to a higher than normal voltage, we get a higher than normal current and, as a result, a higher than normal _____ _____.

[92]

heat loss / power loss

---

[93]

If this voltage becomes much higher, say over 10% above the voltage rating of the device, the heat loss may begin to damage or weaken the wires inside of the device. We then say that _____ _____ shortens the life of an electrical device.

[93]

high voltage

---

[94]

If we were to operate a 120-volt bulb at 175 volts, we would have a high heat loss and would _____ the life of the bulb.

[94]

shorten

---

[95]

If we have an electrical device which seems to fail or burn out much sooner than normal, one cause could be _____ _____.

[95]

high voltage

[46] Refer to Panel III-F, Part (c). The loads A and B consume their share of the source voltage, based on their current and _____. The total of these two voltage drops cannot exceed the _____ voltage.	[46] resistance source / supply
[47] The current through each load in a series circuit will be the same but the _____ _____ across each load may be different, depending on the _____ of the load.	[47] voltage drop resistance
[48] In a series circuit, then, each load has a certain voltage drop which is equal to the product of its own resistance and the current through it, or V = _____. (Formula)	[48] $V = IR$
[49] The voltage drop in a circuit must add up to the supply voltage of the circuit regardless of how many voltage drops there are. Therefore, E always equals the _____ of all the voltage drops (V).	[49] sum
[50] The summation of all of the individual voltage drops in a series circuit equals the _____ voltage.	[50] supply / source

VI - HIGH AND LOW VOLTAGE 112

[86]

We stated before that power is developed by electrical devices in three basic ways: light, heat, and motion. To be more correct we should say that regardless of the type of power developed, there is always some heat present. Therefore, although a motor develops motion and a bulb develops light, each develops a certain amount of _____.

[86]
heat

[87]

The amount of heat which an electrical device develops depends on the amount of current through the wires in it. As the current increases, the heat _____. Why?

[87]
increases

Heat = power = I^2R, IE.

[88]

The heat which some devices develop is not always desirable. Motors are intended to convert power to motion; bulbs, power to light. However, since heat is a form of energy and equals I^2R or IE, any electrical device must have some heat. Because it is the internal wires of a device which have the resistance and current flow, it is in these wires that the _____ is produced.

[88]
heat

[89]

Since we do not want heat in electrical devices such as motors or transformers, we call the heating of the wires in these other devices power loss or heat loss. Why?

[89]
Although we don't want this heat, it is still power and we must pay for its use.

[90]

The resistance of an electrical device is fixed in manufacture. Therefore, the current through the device is determined by the _____ supplied to it.

[90]
voltage

[51]

The circuits below show pictorially what we mean. Note that the sum of the voltage drops (V) equals the supply voltage.

In both circuits, E = 120 volts and the total V equals _____ volts.

[51]

120

[52]

Refer to Panel III-G. Answer the following, using symbols from this panel.

Total resistance or R_T = ___ + ___ + ___.

E = ___ + ___ + ___.

[52]

$R_T = R_1 + R_2 + R_3$

$E = V_1 + V_2 + V_3$

[53]

Refer to same panel. The current through R_1 = current through ____ and ____ and also the current supplied from the source.

[53]

R_2 and R_3

[54]

The voltage drops V_1, V_2 and V_3 [will/won't/may] be equal, depending on their _____. The total of these three voltage drops will be [greater than/less than/equal to] E.

[54]

may

resistance

equal to

[55]

Any time current flows through a load or resistance there will be a _____ _____ across the load.

[55]

voltage drop

[81]

By the manner in which a customer's service wires are connected to the power company's wires, he can pick a voltage of either _____ or _____ volts.

[81] 120
240

[82]

Electrical appliances are always marked as to what voltage they should use. A bulb marked "100-watt 120-volt" should be connected to the _____-volt wires of the power company.

[82] 120

[83]

If a 120-volt bulb is connected to the 240-volt wires, the bulb would be subjected to _____ voltage.

[83] high

[84]

Since current flow through a device is largely determined by the voltage supplied to the device, doubling the voltage will cause the current to _____.

[84] double / increase

[85]

As the current increases, the power used by the device will _____. Why? _____.

[85] increase
P = EI, I²R, IV

[56]

As a summary of all we have just learned, refer to Panel III-H and, after reviewing it, refer to it as needed to solve the following. the following.

(1) E = _____   (4) I_{Total} = _____

(2) R_{Total} = _____   (5) I_{Motor} = _____

(3) V_{Motor} = _____   (6) V_{Total} = _____

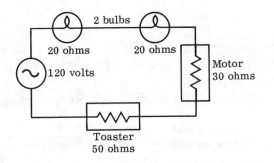

[56]

(1) 120 volts   (4) 1 amp

(2) 120 ohms   (5) 1 amp

(3) 30 volts   (6) 120 volts

[76]

In general, the customer's low-voltage problems are not a result of low supply voltage from the utility company, since considerable amounts of money are spent on equipment designed to furnish the customer with proper _____ voltages.

supply / source

[77]

We have spent considerable time mentioning the problems caused by low voltage. It is also possible to have a high-voltage condition. In general if a device is operated at a voltage over 10% above its rating, this would be called _____ voltage. Again we should emphasize that voltage for some devices would be considered too high at a value much less than 10% above their rating, whereas some devices could operate at a voltage greater than 10% above their rating.

high

[78]

First we will point out that on most utility companies' systems, high voltage to customers is very seldom a problem. The voltage delivered to customers is regulated by elaborate and expensive equipment connected to the power company's system and is maintained at a proper level. This equipment protects the system from _____ voltage.

[Note: High voltage is sometimes referred to as "overvoltage."]

high

[79]

The customer generally does not cause a high-voltage condition because he cannot increase his voltage higher than that supplied by the power company. A customer can overload his circuits and get low voltage but he cannot _____ high-voltage problems.

cause

[80]

However, a customer can cause a high-voltage condition by making a mistake. The power company supplies to most customers two voltages commonly called 120 and 240 volts and written 120/240.

IV Parallel Circuits

[1] We said before that there are two kinds of basic circuits: series and parallel. We have studied the _____ circuit, in which the loads are connected in a(n) _____ arrangement.	**[1]** series series / end-to-end
[2] We will now discuss <u>parallel</u> circuits. In a parallel circuit the loads are not connected in series but are connected so that each load offers the current a different path in which to flow. Therefore, a circuit in which the current has more than one path is a _____ circuit. (See Panel IV-A.)	**[2]** parallel
[3] The parallel circuit is the type of circuit we have in our homes for distribution of electricity for our needs. When we connect appliances or loads to our house wiring, we are giving the electricity another p_____.	**[3]** path
[4] Since, when we add appliances to the circuits in our homes, we add additional paths for the current, we say that we have _____ circuits in our homes.	**[4]** parallel
[5] In a series circuit, the current has only _____ path. In a parallel circuit the current has _____ paths.	**[5]** one many / several

[71]

We stated in the study of parallel circuits that our power system is large enough so that we do not have to worry about our supply voltage being low. This was done to prevent having to bring in the term "voltage drop" at that time.

[72]

In actual practice, however, it is possible that the voltage supplied to a customer is lower than it should be. Each customer's line is connected to the power system by conductors. Since there is, naturally, current flowing in these conductors, they will have some _____.

[72] voltage drop

[73]

If the current in these wires becomes excessive and exceeds the current rating of the conductors, the voltage drop of these conductors [will/will not] result in low voltage at the load.

[73] will

[74]

Every time the customer adds an appliance or other electrical device, he requires more supply current which must come through these wires. Therefore, if he adds too many appliances, the supply conductors may develop a large _____.

[74] voltage drop

[75]

The only way to solve this problem, since the customer cannot reduce his current demand and still use the appliances he wants, is to put in _____ supply wires or service entrance (or both).

[75] larger

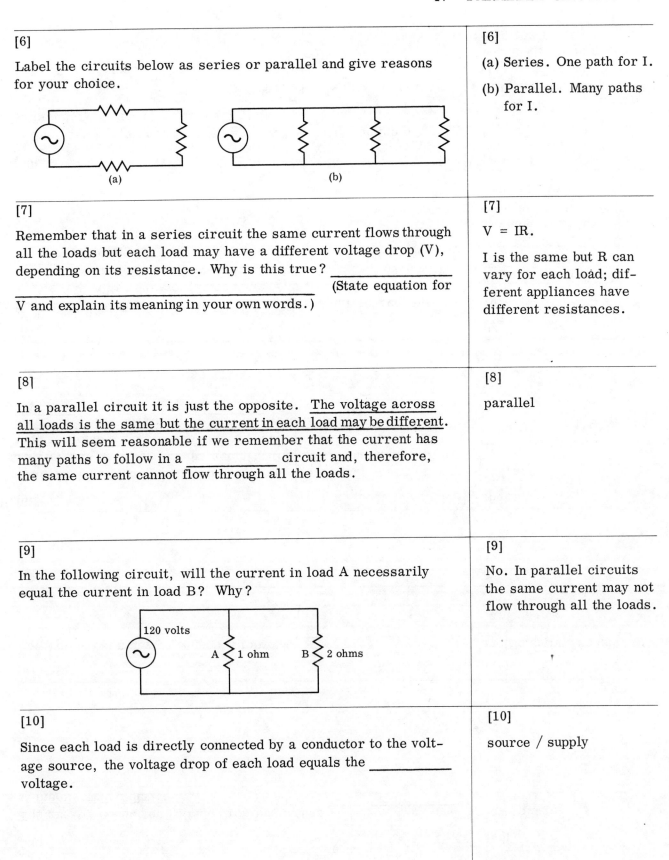

[6] Label the circuits below as series or parallel and give reasons for your choice.	**[6]** (a) Series. One path for I. (b) Parallel. Many paths for I.
[7] Remember that in a series circuit the same current flows through all the loads but each load may have a different voltage drop (V), depending on its resistance. Why is this true? _____ _____ (State equation for V and explain its meaning in your own words.)	**[7]** $V = IR$. I is the same but R can vary for each load; different appliances have different resistances.
[8] In a parallel circuit it is just the opposite. <u>The voltage across all loads is the same but the current in each load may be different</u>. This will seem reasonable if we remember that the current has many paths to follow in a _____ circuit and, therefore, the same current cannot flow through all the loads.	**[8]** parallel
[9] In the following circuit, will the current in load A necessarily equal the current in load B? Why?	**[9]** No. In parallel circuits the same current may not flow through all the loads.
[10] Since each load is directly connected by a conductor to the voltage source, the voltage drop of each load equals the _____ voltage.	**[10]** source / supply

[66]

Too great a conductor voltage drop may cause ____ _____ at the electrical device.

[66]

low voltage

---

[67]

Low voltage at a device causes it to work _____ .

[67]

inefficiently / poorly

---

[68]

Since an appliance is designed to operate at a voltage which is 10% higher or lower than its actual rating, over what range of voltage would a 115-volt appliance operate properly? Minimum = _____ volts; maximum = _____ volts.

[68]

103.5

126.5

---

[69]

If an electrical device is rated at 115 volts and our source voltage is 120 volts, our conductor voltage drop can be about _____ volts before we have voltage so low that the device will not operate properly.

[69]

16.5

Explanation: 115 volts − 10% = 103.5 volts; 120 − 103.5 = 16.5 volts. Device can operate at 103.5 volts, a drop of 16.5 from the source voltage.

---

[70]

In the circuit below do we have "low" or "normal" voltage for satisfactory operation of the bulb?

Conductor V = 5 volts
120 volts

Bulb
100 watts
115 volts

Conductor V = 5 volts

[70]

Normal

Explanation: 120 − (5 + 5) = 110 volts available; 115 − (115 × 10%) = 103.5 volts required; 110 is greater than 103.5.

[11]

In a parallel circuit, then, the voltage drop of each load equals
the _____ _____, and the voltage drops of all loads
in the circuit are _____ _____.

[11]

source voltage

the same / equal

---

[12]

Although the current in each load of a parallel circuit may be
_____, the voltage drop across each load is the same.

[12]

different / unequal

---

[13]

Not only is the _____ _____ across each load the same
in a parallel circuit, but it will also equal the source voltage of
the circuit.

[13]

voltage drop

---

[14]

Observe the circuit below, which is a [series/parallel] circuit,
and answer the following questions.

(1) What is the supply voltage E? _____ volts.
What is the voltage drop V across each load:

(2) V of A = _____ volts?

(3) V of B = _____ volts?

(4) V of C = _____ volts?

[14]

parallel

(1) 100

(2) 100

(3) 100

(4) 100

---

[15]

Recall that our conductors have zero resistance, and we will have
zero voltage drop in the conductor (V = IR). Then the voltage
drop will occur only at the load.

[15]

[61] low voltage

[61] The significance of the second statement from Frame 60 means that appliances rated at 120 volts will generally operate from about 108 to 132 volts (10% of 120 = 12), although in some cases this may slightly reduce the efficiency of their operation. There- fore, with this in mind, an appliance would have _____ if the voltage at the appliance went below 108 volts.

[62] Note: The general operating procedure of most power companies is to maintain voltage at a level for the proper operation of cus- tomers' equipment. In almost all cases, this is well within this 10% allowable variation just discussed.

In specific cases of voltage problems, you should consult the responsible person in your area.

[63] 10%

[63] This 10% figure is only an approximate value but, in general, elec- trical devices will operate properly if they are in this range of above or below their voltage rating.

[64] (1) current (2) resistance

[64] In summary, we have found that electrical devices will not oper- ate properly with a low-voltage condition. This condition is caused by

(1) excessive _____ flowing in the conductor, or

(2) a conductor which has too high a _____.

[65] resistance voltage drop equals current × resistance.

[65] Excessive current in a conductor or too much conductor will cause a large voltage drop along the con- ductor because _____ (Own words)

[16]

The voltage drop in a circuit occurs across the _____, not in the _____.

[16]

load

conductors

---

[17]

The voltage drop between A and B in the circuit below is ___; the voltage drop between B and E is ____; the voltage drop between B and C is ___, while between C and D the voltage drop is ____.

120 volts — A, B, C (top), F, E, D (bottom)

[17]

0

120

0

120

---

[18]

Since there is no voltage drop along the conductor ABC or DEF, the voltage between these conductors is at every point equal. The voltage from A to F, then, equals the voltage from B to E and ___ to ___.

[18]

C to D

---

[19]

In a parallel circuit, therefore, the same voltage drop appears across each l_____, regardless of the number of loads connected.

[19]

load

---

[20]

Since the only voltage available to the circuit is from the supply voltage, the load voltages for all loads are equal to each other and also equal to the _____ _____.

[20]

supply voltage

[56]

Therefore, if there is an excessive voltage drop which is causing low voltage, the present fuse in the circuit is apparently [larger/smaller] than it should be.

larger   [56]

---

[57]

The fuse in a circuit should, for safety, be [equal to/smaller than/larger than] the current rating of the conductor. (Pick two.) This will protect the conductor against overheating and also prevent _____ _____.

equal to / smaller than
low voltage   [57]

---

[58]

So we see that V = IR, voltage drop in a conductor, depends on

(1) _____ (in conductor),

(2) _____ (of conductor).

(1) current
(2) resistance   [58]

---

[59]

We have also seen that the resistance of the conductor depends on

(1) conductor _____,

(2) conductor _____.

size
length   [59]

---

[60]

It may seem very hard to get the exact size of conductor for a circuit to avoid too much voltage drop and still not require large bulky conductors. It isn't, because (1) in many cases, the conductor is short enough so that its total resistance and voltage drop are insignificant, (2) the rating of electrical devices is not an exact requirement and generally allows for deviations of + or − 10% before their operation significantly suffers, although in certain situations this much voltage deviation cannot be allowed.

[60]

[21]

In a parallel circuit, the voltage drop across each load is [the same/different] and equal to the _____ _____.

[21]

the same

supply voltage

[22]

Since the voltage across each load is the same in a parallel circuit, we can find the current in each load if we know the load resistance. The equation we should use to find this current is

$I = \dfrac{V}{-}$.

[22]

R

[23]

Refer to the circuit below.

The current load in A is found as follows:

$I_A = \dfrac{100}{2} = 50$ amp.

100 volts

2 ohms   A   4 ohms   B

Find the current in load B. $I_B =$ _____.

[23]

25 amp

Explanation:

$I_B = \dfrac{100}{4} = 25$ amp.

[24]

In the circuit in the previous frame, when load A has 50 amp and load B has 25 amp, our total current from the source must be 50 + 25 = 75 amp. We can then say that in a parallel circuit, the total current from the source must be equal to the sum of the currents in each _____.

[24]

load

[25]

We have seen that the same current does not flow through each load in a _____ circuit.

[25]

parallel

[51]

If the voltage source is _____ the sum of the voltages required by the individual loads, we will have a low-voltage problem.

[51]

less than

[52]

Two ways to reduce the voltage drop of a conductor in a circuit are to _____ the size of the conductor or _____ the length of the conductor, if possible.

[52]

increase

decrease

[53]

Voltage drop in a conductor may also be caused by a high current in the conductor since $V = IR$. Therefore, if our current exceeds the rating of the conductor, we can expect an excessive voltage drop in the conductor and a _____ - _____ problem.

[53]

low-voltage

[54]

If our current is excessive, we can do two things:

(1) _____ the current by adding another circuit to share in carrying the current,

(2) put in a _____ conductor which would be rated to carry the required current.

[54]

(1) decrease

(2) larger

[55]

In general, if the conductor is properly fused (fuse rating equals current rating of conductor), it will not allow excessive current to flow and this will prevent an excessive _____ _____ in the conductor.

[55]

voltage drop / current

flow / load

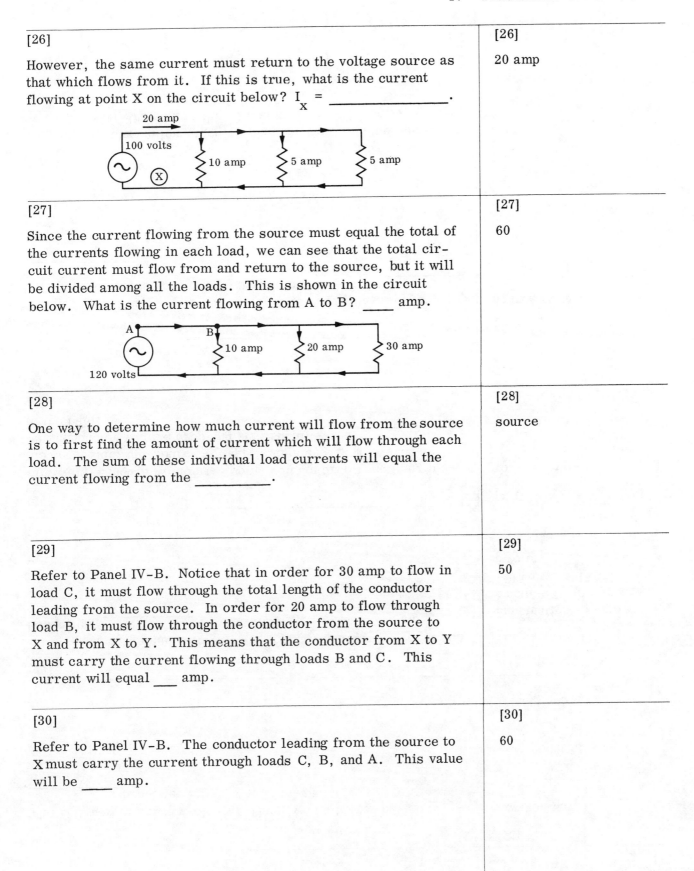

[26]

However, the same current must return to the voltage source as that which flows from it. If this is true, what is the current flowing at point X on the circuit below? I$_X$ = _____.

[26]

20 amp

[27]

Since the current flowing from the source must equal the total of the currents flowing in each load, we can see that the total circuit current must flow from and return to the source, but it will be divided among all the loads. This is shown in the circuit below. What is the current flowing from A to B? ____ amp.

[27]

60

[28]

One way to determine how much current will flow from the source is to first find the amount of current which will flow through each load. The sum of these individual load currents will equal the current flowing from the _____.

[28]

source

[29]

Refer to Panel IV-B. Notice that in order for 30 amp to flow in load C, it must flow through the total length of the conductor leading from the source. In order for 20 amp to flow through load B, it must flow through the conductor from the source to X and from X to Y. This means that the conductor from X to Y must carry the current flowing through loads B and C. This current will equal ___ amp.

[29]

50

[30]

Refer to Panel IV-B. The conductor leading from the source to X must carry the current through loads C, B, and A. This value will be ____ amp.

[30]

60

[46]

The bulb now has rated voltage. Thus, we no longer have a _____ problem.

[46] low-voltage

[47]

Do we still have some voltage drop in our conductor? If so, _____ how much?

[47] Yes. 5 volts

Explanation: 5 amp × 1 ohm = 5 volts or 120 volts – 115 volts = 5 volts.

[48]

However, although we do have some voltage drop remaining in the conductor, our source voltage is high enough to overcome this conductor voltage drop and still supply rated _____ to the bulb.

[48] voltage / power

[49]

Another way to say this is that the voltage requirement of the bulb plus the conductor voltage drop do not exceed the _____.

[49] source voltage

[50]

The rule we must follow, then, is that the source voltage must be equal to the sum of the voltage drop of our load(s) and _____ conductor(s).

[50] equal to

[31]	[31]
Refer to Panel IV-B. To summarize, then, the total circuit current will flow in the conductor connecting the first load with the source. At the first load junction (X), the current in the main conductor will [increase/decrease] by an amount equal to the first load current. At each succeeding load, the current in the main conductor will decrease by the amount which then flows through that _____ . Notice how this occurs in the circuit.	decrease load

[32]	[32]
	(1) 30
	(2) 10
	(3) 20
	(4) 30
In the above circuit, 10 amp flow through each load. What current (in amperes) will flow in the conductor	
(1) from A to 1? ____ (3) from 5 to 6? ____	
(2) from 2 to 3? ____ (4) from 6 to B? ____	

[33]	[33]
In parallel circuits the conductors nearest the source will carry [more/less] current than the conductor connecting the last load to the circuit.	more

[34]	[34]
In parallel circuits, the total current supplied from the source will be equal to the sum of the currents in each _____ .	load

[35]	[35]
In a parallel circuit, then, the more loads we connect to the circuit, the more current will flow from the _____ .	source

[41] This would probably decrease the conductor's resistance and, thus, voltage drop sufficiently but may not be practical since the bulb may actually be located 500 ft from the voltage source. Because the length of the conductor we need cannot be normally changed, we usually have only one choice to reduce the conductor voltage drop. We should _____.

(Own words)

[41] increase the size of the conductor

[42] Let us see what increasing the size of the conductor would do. A #10 copper conductor has only one ohm per 1000 ft. (Wire sizes get larger as the wire size number gets smaller.) With the same circuit as in Panel VI-C we will [increase/decrease] the voltage drop of the conductor if we use a #10 conductor instead of #20.

[42] decrease

[43] Refer to Panel VI-D. The total circuit resistance is ____ ohms.

[43] 24

[44] Now determine the circuit current from Panel VI-D.

I = _____

[44] 5 amp

Explanation: $I = \dfrac{E}{R} = \dfrac{120}{24} = 5$ amp.

[45] Refer to Panel VI-D. Our bulb now has its rated current. What is the voltage across the bulb? V = _____

[45] 115 volts

Explanation: $V = IR =$ 5 amp × 23 ohms = 115 volts.

[36]

The source will supply ____ amp when switch A is closed, ____ amp when B and A close, and ____ amp when A, B, and C close.

---

[36]

12 (Remember that I = E/R and R = 10, E = 120).

24

36

---

[37]

As we connect more loads to the parallel circuit and the current from the source increases, the current in the conductors near the source will also _____ .

---

[37]

increase

---

[38]

The basic rule we want to obtain is: the current into a junction (points X and Y below, for example) of conductors must be equal to the current out of the junction on the other side. For example, when (a) is true, what is the current flowing from the junction on conductor 1 in (b)? _____ .

---

[38]

20 amp

Explanation: 50 = 30 + 20.

---

[39]

Refer to the circuit in Panel IV–B and explain how the current is distributed at junctions M and N. (That is, what is the value in amperes of the current flowing into and out of these junctions?)

   Junction M: Current flowing in is ____ + ____ = ____

            Current flowing out = ____

   Junction N: Current flowing in is ____ + ____ = ____

            Current flowing out = ____

---

[39]

20 + 30 = 50

50

10 + 50 = 60

60

---

[40]

Answer the questions for the circuit below. (Recall I = E/R.)

(1) I in load A = ____ amp.
(2) I in load B = ____ amp.

   I flowing in
(3) Junction 1 = ____ amp.
(4) Junction 2 = ____ amp.
(5) Junction 3 = ____ amp.

---

[40]

(1) $I_A = 10$

(2) $I_B = 5$

(3) $I_1 = 15$

(4) $I_2 = 5$

(5) $I_3 = 15$

[36]

Refer to Panel VI-B. The bulb is required to have 115 volts for proper operation. We have seen that our circuit will allow it to have only about 84 volts across it. We therefore have _____.

[36]

low voltage

[37]

Refer to Panel VI-B. Our source was 120 volts, and we found that our bulb had only 84 volts across it. Where do we drop the rest of the voltage? _____.

[37]

Along the conductors.

[38]

To summarize the entire picture of voltage drop, refer to Panel VI-C.

[38]

[39]

To remedy this specific low-voltage problem, there are two possibilities:

(1) reduce the length of the conductors,
(2) increase the _____ of the conductors.

[39]

size

[40]

If 1000 ft of conductor gives us 10 ohms, 500 ft would be only 5 ohms and 10 ft would be only _____.

[40]

0.1 ohm / $\frac{1}{10}$ ohm

[41]

Another way to determine the total circuit current that must be supplied from our source would be to use the standard equation I = E/R. To do this we must first determine our total circuit resistance.

[41]

---

[42]

The equation I = E/R states that total supplied current equals supply voltage divided by total circuit _____.

[42]

resistance

---

[43]

In the series circuit our total resistance was equal to the _____ of all the resistances or loads and became [larger/smaller] as we added loads to the circuit.

[43]

sum

larger

---

[44]

In a parallel circuit, the total R is determined as follows:

$$1/R_{Total} = 1/R_1 + 1/R_2 + 1/R_3 + 1/R_4 + \cdots$$

Or, the reciprocal of total circuit resistance is equal to the _____ of the _____ of all the circuit _____.

[Note: Reciprocal of 2 = 1/2, 9 = 1/9, R = 1/R, etc.]

[44]

sum

reciprocals

resistances

---

[45]

"Reciprocal" is a mathematical term. To obtain a reciprocal, change the number you are using to a fraction, then invert it (turn it upside down). Example:

Reciprocal of
  2 (fraction 2/1) = 1/2,
  9 (fraction 9/1) = 1/9,
  2/3 (fraction 2/3) = 3/2.

[45]

[31]

20

For example: The resistance of #20 copper conductor is 10 ohms per 1000 ft. Therefore, if we have 2000 ft of #20 copper conductor, we would have ____ ohms of conductor resistance.

[32]

115

Refer to the drawing in Panel VI-B. In order for the bulb to operate properly, the circuit requires 5 amp of current and ____ volts.

[33]

33 ohms

Explanation: 10 ohms + 23 ohms = 33 ohms.

Since in VI-B we have 1000 ft of #20 conductor (500 ft out and 500 ft back), we have a total of 10 ohms of resistance in the conductor. [Note: From tables it may be found that #20 wire has a resistance of about 10 ohms per 1000-ft length.] This conductor resistance is treated as being in series with the bulb. What is the total circuit resistance?

[34]

3.63 amp

Explanation: $I = \dfrac{E}{R} = \dfrac{120}{33} = 3.63$ amp.

Refer to Panel VI-B. If the total circuit resistance is 33 ohms, what is the circuit current?   I = ____

[35]

83.5 volts (approx.)

Explanation: 83.5 volts = 3.63 amp × 23 ohms.

We therefore have 3.63 amp flowing through the bulb. Since the bulb resistance is 23 ohms, what is the voltage across the bulb?

[46]

The reciprocal of 3/5 is

(1) $\frac{6}{10}$ (2) $1\frac{2}{5}$ (3) $\frac{5}{3}$

Check one, look at the answer, then follow the instructions given.

[46]

(1) Go to Frame 47.

(2) Go to Frame 47.

(3) Go to Frame 54.

---

[47]

Somewhere you have been misled. You were asked to give the reciprocal of 3/5. Your answer should have been 5/3. Remember that in Frame 45 you were told that to find a reciprocal means to invert (turn upside down) the fraction? Thus, the reciprocal of 3/5 becomes 5/3. What is the reciprocal of 1/4?

_____

[47]

$\frac{4}{1}$

---

[48]

Absolutely! All we do is invert the fraction number, making the top number (the numerator) in the old fraction become the bottom number (the denominator) in the new fraction. Thus, if we were to write the reciprocal of 5/8, where would the 5 be in our new fraction? _____

[48]

In the denominator /
in the bottom number.

---

[49]

Correct. To complete the rest of the reciprocal, the bottom number (the denominator) will become the _____ number (the numerator) in the new fraction. Thus, in our previous example of 5/8, the 8 would become the _____ _____ in our new fraction.

[49]

top

top number / numerator

---

[50]

Therefore, the complete reciprocal of the fraction 5/8 is _____.

[50]

8/5

[26]

If a motor runs too slowly, a possible cause could be _____ _____ in its circuit.

| [26] |
| low voltage |

---

[27]

To visualize this low voltage, consider the circuit in Panel VI-A. The motor in the panel requires ____ volts and ____ amp to operate properly.

[27]

115

10

---

[28]

In the circuit in Panel VI-Λ, we could allow the conductor a voltage drop of ___ volts and still allow the motor to have rated voltage.

[28]

5

Explanation: 120 volts source − 115 volts required = 5 volts extra.

---

[29]

Motor rating: 118 volts
10 amp

120 volts   Motor

How much voltage drop could we allow in this conductor? _____

[29]

2 volts

---

[30]

The resistance of a conductor is rated in ohms per foot, meaning that the total resistance of the conductor is spread out over the entire length of the _____.

[30]

conductor

[51]

Let's see if all this is clear now. What would the reciprocals of the following numbers of fractions be?

    (1) 5/6, reciprocal = _____

    (2) 3/2, reciprocal = _____

    (3) 2,   reciprocal = _____

[51]

(1) 6/5

(2) 2/3

(3) 1/2

If you did not give this as the answer, go to Frame 52. If this was your answer, skip to Frame 54.)

[52]

All right, part (3) of Frame 51 gave you some trouble. The easiest way to get this kind is to remember that any whole number, 2, 3, 6, 12, etc., is understood to have a denominator of 1 (the bottom number). Thus, when you invert the number 2/1, the reciprocal becomes 1/2. What is the reciprocal of 10?

[52]

1/10

[53]

Fine. Let's review once more. To find the reciprocal of any number, invert its fraction form, remembering there is always a denominator of 1 understood in each whole number. Find the reciprocals of the following.

    (1) 7/3   (2) 1/5   (3) 8

[53]

(1) $\dfrac{3}{7}$ (2) $\dfrac{5}{1}$, 5   (3) $\dfrac{1}{8}$

(You still have problems? Return to Frame 45 and review reciprocals.)

[54]

An example will best describe this method of determining **parallel circuit resistance.** The total resistance of the following circuit is found as follows: $1/R_T = 1/4 + 1/3 + 1/6 = 3/12 + 4/12 + 2/12$. Therefore, $1/R_T = 9/12$. $R_T = 12/9 = 4/3$ or $1\frac{1}{3}$ ohms.

[54]

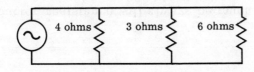

[55]

What would be the total resistance (R_T) in the circuit below? Check one, look at the answer, then follow the instruction given.

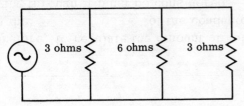

    (1) 10/12 or 5/6 ohms

    (2) 12 ohms

    (3) 12/10, 6/5, or

         $1\frac{1}{5}$ ohms

[55]

(1) Go to Frame 56.

(2) Go to Frame 56.

(3) Go to Frame 68.

[21]

Therefore, to regulate the amount of voltage drop in a conductor, both the _____ of the conductor and the _____ current flowing through it must be considered.

[21] size / resistance / current

[22]

The actual conductor resistance will always be much [less/more] than the resistance from the connected loads.

[22] less

[23]

Even though the resistance of a conductor is very low when compared with the resistance of the connected loads, it must be considered when trying to determine the proper size of the con- ductor required to prevent large _____ _____ in the conductor.

[23] voltage drop

[24]

We must consider the voltage drop of a conductor because all electrical devices are made to operate at a certain _____.

[24] voltage

[25]

If electrical appliances are operated at voltages less than their rated voltage, they will operate _____.

(Own words)

[25] improperly / inefficiently / below standard

[56]	**[56]**
You're having a problem with arithmetic and its perfectly understandable. You see, in order to solve Frame 54, we must substitute the resistance values in the formula given in Frame 44 ($1/R_T = 1/R_1 + 1/R_2 + \cdots$). Thus, in our problem, $1/R_T = 1/4 + 1/6 + 1/\underline{}$.	3
[57]	**[57]**
Our complete arithmetic problem is, then, $1/R_T = 1/4 + 1/6 + 1/3$. To solve this problem, which is simply addition of fractions, we must first find the <u>smallest</u> common _____ of these fractions.	denominator
[58]	**[58]**
By <u>smallest</u> common denominator (sometimes called LCD or least common denominator), we mean "the _____ number each of the fraction's denominators will divide into evenly."	smallest
[59]	**[59]**
While there is a system for finding this LCD, it can be quite complicated. Many times, the easiest way to find it is by trial and error. What is the LCD of 1/4, 1/6, and 1/3? ____	12 Explanation: 24 or 36 would be common denominators but they would not be the smallest.
[60]	**[60]**
When we have found this LCD, we change each of the fractions in our problem to the equivalent fraction with that common denominator. Thus, $1/4 = 3/12$, $1/6 = 2/12$, and $1/3 = \underline{}/12$.	4

[16]
By reducing either the conductor resistance or current, we can reduce the _____ _____ of the conductor.

[16] voltage drop

[17]
Since the size of the wire determines the amount of resistance it has, we can reduce the wire resistance by _____ its size.

[17] increasing

[18]
Naturally, we cannot continue to reduce the resistance of the wire by increasing the wire size without limit, or it will become too bulky and expensive. Therefore, we must pick a proper and reasonable wire size which will give us a small enough amount of _____ in our circuit to prevent excessive voltage drop.

[18] resistance

[19]
Since our conductor resistance is somewhat set by the size limitations of the conductor, we must also consider the _____ flowing in the conductor, to control our conductor voltage drop.

[19] current

[20]
In any given conductor which has a fixed _____, an increase in current will cause an increased _____ in that conductor.

[20] resistance
     voltage drop

[61]

Our next step, then, is just to add the 3 fractions with the same denominators. Thus, $3/12 + 2/12 + 4/12 = $ _____ .

[61]

9/12

(Did you say 9/36? Remember, we add only the numerators. This is like adding 4 eggs + 2 eggs + 3 eggs = 9 eggs.)

[62]

Now our formula should look like this: $1/R_T = 9/12$. But we want to know what R_T equals, not $1/R_T$. So all we do is turn (invert) $1/R_T$ into R_T. When we do this, we must also invert 9/12. What do we get when we invert 9/12? _____ .

[62]

$\frac{12}{9}$ / $\frac{4}{3}$ / $1\frac{1}{3}$

[63]

Fine, if you said 4/3 or $1\frac{1}{3}$; you were way ahead of us. You remembered that to be absolutely correct we should reduce our answer to its lowest possible terms. Since in the fraction 12/9 both the numerator and denominator can be divided by the number ___, 12/9 does reduce to _____ .

[63]

3

4/3

[64]

Now, 4/3 is called an improper fraction, and even though it is not incorrect to leave an answer as an improper fraction, many people commonly change the improper fraction 4/3 to a mixed fraction, which in this case is _____ .

[64]

$1\frac{1}{3}$

[65]

Therefore, our final answer to the question in Frame 54 would be $R_T = $ _____ ohms.

[65]

$1\frac{1}{3}$

[11] no / zero resistance	[11] We have stated previously that a perfect conductor has no / zero resistance but that in actual practice all conductors have some _____ , even though it may be small.
[12] I × R	[12] As a matter of review, voltage drop in a wire is found by the formula V = _____ × _____ .
[13] voltage	[13] Since conductors do have resistance when current flows through them, we have a _____ drop in the conductor.
[14] low	[14] This conductor voltage drop, just like the heating of the conductor, is undesirable and helps us in no way. In other words, it is a voltage loss which could cause [high/low] voltage at our load.
[15] R	[15] Since V of the conductor equals the value of IR, there are two ways we can keep this voltage drop to a value low enough so that it is not a problem. We can reduce V by reducing either _____ or I.

[66] Just to check ourselves, let's see what the total resistance in the circuit below would be. R_T = _____ ohms *(circuit diagram: AC source with three parallel resistors — 6 ohms, 8 ohms, 3 ohms)*	[66] $1\frac{3}{5}$ ohms Explanation: $R_T = 24/15 = 8/5 =$ $1\frac{3}{5}$ ohms.
[67] If you agreed with any of the answers given for Frame 66, you're ready to go ahead to Frame 68. If your answer differed from those given, we suggest you return to Frame 57 and review this problem.	[67]
[68] Refer to Panel IV-C. If the circuit in Panel IV-C were broken at 1 and 2, making the circuit contain only branch ABGH, what would be the resistance of the circuit? R = _____.	[68] 2 ohms
[69] Determine the total resistance of the entire circuit in Panel IV-C. R_T = _____.	[69] $1\frac{1}{11}$ ohms Explanation: $1/R_T = 1/2 + 1/4 + 1/6 = 6/12 + 3/12 + 2/12$; $1/R_T = 11/12$; $R_T = \frac{12}{11} = 1\frac{1}{11}$ ohms.
[70] In the illustration in Panel IV-C, which circuit had the <u>smaller</u> resistance, the parallel circuit with all of the loads connected or the circuit ABGH just discussed in Frame 68?	[70] Circuit with several loads, the parallel circuit.

[6]

If we have less than this required voltage, let us see what happens. If we have only 50 volts, what current will we have? (Remember that the bulb R will not change.)

I = _____

[6]

0.5 or 1/2 amp

Explanation: $I = \dfrac{V}{R} = \dfrac{50 \text{ volts}}{100 \text{ ohms}} = 0.5$ amp.

[7]

With a current of 0.5 amp flowing through the bulb and with 50 volts applied to the bulb, how much power is required by the bulb? _____

[7]

25 watts

Explanation: $P = IV = 0.5 \times 50$.

[8]

Therefore, although the bulb is rated at 100 watts, we can have this wattage only if we have a voltage of 100 volts. As our voltage decreases, the power available to the bulb also _____ and we have an inefficient bulb with less light.

[8]

decreases

[9]

The power available to a device decreases and the usefulness of our device decreases if we have a l_____ in our circuit.

[9]

low voltage

[10]

In order for an electrical device to operate properly, it should be used in a circuit that supplies the voltage for which the device is rated. If the voltage is below this value, we say we have _____ and the device [will/will not] operate properly.

[10]

low voltage

will not

[71]

Refer to Panel IV-C. If now we would use the circuit ABCFGH which would contain the 2- and 4-ohm resistances, what would be the total circuit resistance?

[71]

R_T = 4/3 = $1\frac{1}{3}$ ohms.

Explanation:
$1/R_T$ = 1/2 + 1/4 = 2/4 + 1/4 = 3/4; R_T = 4/3.

---

[72]

Refer to Panel IV-C for the next four frames. The circuit with only the 2-ohm resistance had a total resistance of 2 ohms; the circuit with the 2- and 4-ohm resistances had $1\frac{1}{3}$ ohms and the circuit with all 3 resistances had $1\frac{1}{11}$ ohms resistance.   From these figures we can summarize that in a parallel circuit, as we add more loads in parallel, the total circuit resistance [increases/decreases].

[72]

decreases

---

[73]

Using the relationship I = V/R, find the current in each load and determine the total source current.

     (1) I (2 ohm) = _____ amp

     (2) I (4 ohm) = _____ amp

     (3) I (6 ohm) = _____ amp

     (4) Total current = _____ amp

[73]

    I = V/R =

(1)   120/2 = 60

(2)   120/4 = 30

(3)   120/6 = 20

(4)   Total = 110 amp

---

[74]

Now determine the total current supplied, using I = E/R. I = _____. (Use the answer for R_T from Frame 69.)

[74]

110 amp

Explanation:

$I = \dfrac{E}{R} = \dfrac{120}{12/11} = 110$ amp.

---

[75]

You got 110 amp for total source current by adding all individual load currents and this same total by finding total circuit R and dividing source voltage by R. We find total source current in a parallel circuit by

    (1) _____
                    (Own words)

    (2) _____
                    (Own words)

[75]

(1) finding current in each load by I = V/R and adding the currents together,

(2) finding total circuit R by $1/R_{Total}$ = $1/R_1$ + $1/R_2$ + $1/R_3$ and using I = E/R for total current.

VI High and Low Voltage

[1]

You may have heard the term "low voltage." Low voltage means voltage which, when applied to our electrical device, is too _____ to allow the device to function properly. [Note: Low voltage is sometimes referred to as "undervoltage."]

low

[2]

If we have a motor designed to operate at 240 volts and the voltage delivered to the motor is [200/280] volts, we say that we have low voltage at the motor.

200

[3]

If we do not supply enough voltage to a device to allow it to operate properly, we have _____ .

low voltage

[4]

Appliances are designed to operate at a certain voltage and are made with a certain set resistance. For example, a bulb rated "100-watt 120-volt" must have 120 volts before it can develop _____ watts of power.

100

[5]

From the power equation $P = IV$, we can determine that a 100-watt bulb designed to operate at 100 volts will have 1 amp of current flowing through it. We find this as follows:

$$P = IV, \qquad 100 = I \times 100, \qquad I = \frac{100}{100} = 1 \text{ amp.}$$

This bulb must then have a resistance of _____ ohms.

100
(Since $I = V/R$, and from Frame 5, I was set as 1 amp and V as 100 volts; then $1 = 100/R$. Therefore, R must be 100.)

[76]

In a parallel circuit, as we add more loads in parallel, the total resistance [decreases/increases] and the total supplied current _____.

[76]

decreases

increases

---

[77]

In a series circuit, as we added loads, our total resistance [increased/decreased] and since I = E/R, the total current _____.

[77]

increased

decreased

---

[78]

Find the total R in the following circuits.

R$_{Total}$ _____   R$_{Total}$ _____

[78]

(a) 6 ohms
Explanation: 2 + 4 = 6.

(b) $1\frac{1}{3}$ ohms
Explanation: $1/R_T = 1/2 + 1/4 = 3/4$; $R_T = 4/3 = 1\frac{1}{3}$ ohms.

---

[79]

In a series circuit the same current will flow through all the loads. In a parallel circuit the _____ across each load will be the same but the _____ through the load will depend on the resistance of the load.

[79]

voltage

current / amperes

---

[80]

Did you notice that when we added loads in the series circuit, the total circuit resistance increased, but when we added loads in the parallel circuit, the total circuit resistance decreased? This was the case, and the following rule will always apply: When you add loads in a series circuit, the total resistance _____; but when you add loads in a parallel circuit, the total circuit resistance _____.

[80]

increases

decreases

[136]

The temperature of the fuse wire is determined by the fuse wire resistance and circuit _____ .

[136]

current / load

[137]

We will not thoroughly investigate the device known as the circuit breaker, but for practical purposes you may consider that it serves the same purpose as a fuse. The difference is that after it "blows," you need only reset it rather than replace it.

[137]

[138]

As a summary for the basic concepts of power, refer to Panel V-H.

[138]

[81]

Since the relationship I = E/R is always true, you would expect the current to [increase/decrease] in a series circuit when you add loads.

[81]

decrease

[82]

Using the same relationship, you would expect the current to increase in a _____ circuit as you add loads to the circuit.

[82]

parallel

[83]

This relationship of current and load may be clearer if you think of it in this way. In the series circuit there is only one path for the current to follow and, when we add loads, we are adding resistance in this one path, which tends to _____ the current flow.

[83]

reduce / decrease

[84]

In the parallel circuit, when we add loads, we are increasing the number of paths in which the _____ can flow, making it easier for more current to flow from our source.

[84]

current

[85]

Therefore, as we add loads in a parallel circuit, we expect our source current to [increase/decrease].

[85]

increase

[131]

Hence, as is evident from the equation $P = i^2R$, we can actually have power loss in a conductor. (We consider it loss since we cannot make good use of it.) Since this power does not develop light or motion, it must be developing _____.

[131] heat

[132]

Thus, we have power loss in our conductors from the development of heat. The resistance of our conductor does not change. As our current increases, the power loss and heat increases — since $P = i^2R$. Therefore, as the current increases, the danger from overload increases and the conductor becomes _____.

[132] hot / heated

[133]

We use this heat to our advantage in the fuse. The material in the fuse wire is such that at a certain temperature it will melt. The fuse wire, like all conductors, has some resistance. There-fore, current flowing through the resistance of the fuse wire gives a power loss in the form of _____.

[133] heat

[134]

As the current in the fuse increases, the temperature of the fuse wire increases and the wire finally melts at the current rating of the fuse. If the fuse size is properly chosen, the fuse wire would melt before the circuit wire becomes _____.

[134] hot / overheated

[135]

Remember that the fuse wire is made of a metal which melts at a temperature such that the circuit wire does not exceed its _____ rating.

[135] current / ampere / maximum

[86]

Since our home wiring is an example of a parallel circuit, as we plug in more appliances (which are loads), our source current will _____.

[86]

increase

[87]

In which of the circuits below will the loads have the same voltage drop? ____   The same current? ____

(a)   (b)

[87]

(b)

(a)

[88]

For a summary of information regarding parallel circuits, refer to Panel IV-E, Parts A and B.

[88]

[89]

As a summary for the discussion of both series and parallel circuits, refer to Panel IV-F, Part A, and supply the missing information. (It may help you to answer the questions if you draw a series and parallel circuit for reference). The correct answers for this panel will be found on Panel IV-F, Part B.

[89]

[126]

Indicate the wire and fuse size that would be required on the following circuits if they contained the devices listed and all were on at the same time. (Refer to Panel V-G for fuse sizes.)

	Wire	Fuse
(1) Toaster 10 amp, clock 1, bulbs 5		
(2) Motor 25 amp, bulbs 2, soldering iron 10		
(3) Bulbs 10 amp, electric clock 1, radio 3		

[126 answer]

	Wire	Fuse
(1)	No. 10	30 amp
(2)	No. 8	40 amp
(3)	No. 14	15 amp

[127]

In order to understand why excessive current will cause a wire to overheat, we must review slightly. We stated previously that conductors have very low resistance, which can normally be neglected. This is not true when dealing with excessive current in conductors, as will be shown in the following frames.

[128]

The resistance of a conductor is determined in part by its length, since small resistances occur all along the conductor, as shown below.

The longer the conductor, for the same wire size, the larger will be its total _____.

[128 answer] resistance

[129]

Power was said to be equal to the following formula: $P = I^2R$. Therefore, when current flows through the conductor, even though the resistance of the conductor is small, there is some heat developed in the _____.

[129 answer] conductor

[130]

This heat is caused by the _____ flowing through the resistance of the conductor.

[130 answer] current

V Power

[1] We will now introduce a new term called <u>power</u>, which will be designated by the letter P and measured in <u>watts</u>. Power is symbolized by the letter ___.	[1] P
[2] Since watts is the name given to power, then a 100-watt bulb will use 100 _____ of power.	[2] watts
[3] When an iron is rated at 100 watts, this tells us how much _____ it will use.	[3] power
[4] The new electrical term we are concerned with is <u>measured</u> in _____, has a <u>symbol</u> of ___, and is called _____.	[4] watts P power
[5] Let's try to visualize just what this power concept actually is. Power is the speed or rate at which we do <u>work</u> or consume energy. When we mow our lawn, for example, we call this _____.	[5] work

[121]

By putting a fuse in our circuit, we limit the amount of current which our circuit can conduct. Therefore, if too many appliances are added to one circuit, causing excessive _____ and, as a result, too much heat, the fuse in that circuit will _____.

[121]
current

melt / blow

---

[122]

Would you use a 40-amp fuse in a circuit with a No. 14 AWG wire which can safely conduct only 15 amp of current? Why?

[122]
No. Circuit could become overloaded and wire hot.

---

[123]

Suppose that we have a circuit wire rated at 15 amp and appliances connected to the circuit which already total 9 amp. We now add an appliance which requires 10 amp. Which of the following fuses would prevent the circuit wire from carrying excessive current and still allow the most efficient and safe use of the circuit wire?

[10 amp/15 amp/20 amp]

[123]
15 amp. (Since wire can safely carry 15 amp.)

---

[124]

If every time you plug in the coffee pot the fuse burns out, and the pot is known to be good, would you

                Yes No

(1) put in a larger fuse,

(2) unplug something else on the circuit
before plugging in the coffee pot,

(3) try another plug for the coffee pot?

[124]
(1) No.
(2) Yes.
(3) Yes.

---

[125]

Panel V-G lists various conductors by size and current rating just as they appear in the National Electrical Code. Refer to this panel to answer the questions in the next frame. (Remember that for economical reasons, the smallest wire size which will safely carry the required current is usually used rather than an over-sized wire.)

[125]

[6] If we mow our lawn very rapidly, we use more _____ than if we mow if very slowly. We must, since we get more tired when we work at a faster rate.	[6] power
[7] If we run up a flight of stairs, it takes more _____ than if we walk.	[7] effort / power
[8] If we hand-pump water very rapidly, it means we are working at a faster rate and therefore using [more/less] power to pump the same amount of water.	[8] more
[9] In this day and age we do very little hand-pumping of water; instead we use motors. These motors still must do the work of pumping the water and therefore supply the _____ for doing the work.	[9] power
[10] Some electric pumps move the same amount of water faster than others; therefore, they use more _____, measured in units of _____.	[10] power watts

[116]

If we pick a wire size for our fuse that will melt at 15 amp, our circuit wire would never be able to conduct more than ___ amp of current.

[116] 15

[117]

The reason that the fuse can protect the circuit wiring is that it is in series with the conductor leading from the source and, if it melts, it will open the circuit just as though a switch had been opened in the circuit. Current cannot flow in an ___ circuit.

[117] open

[118]

In which circuit below will current flow, (a) or (b)? ___

(a) Good fuse   120 volts

(b) 120 volts   Blown fuse (open circuit)

[118] (a)

[119]

When you buy a fuse, it is rated in amperes: 10 amp, 15 amp, 30 amp, etc. This rating means that the fuse will melt if more than that amount of current starts to flow through it. A 30-amp fuse will melt if more than ___ amp flow through it.

[119] 30

[120]

The reason that we need fuses is that we do not always know how much current our electrical appliances will require, and we usually have several in one circuit. Every time we add an appliance to a house circuit, the ___ in that circuit increases.

[120] current

[11]

Since these pump motors are supplying power to pump water, they are rated in _____, which is the unit for measuring electric power.

[11]

watts

[12]

All electric motors do some kind of work for our use (mixers, pumps, grinders, etc.), and since they do this work in a specified length of time or at a certain rate, they all supply _____, which is measured in _____.

[12]

power

watts

[13]

Which motor would be capable of filling a 100-gallon water tank faster, a 1000-watt motor or a 500-watt motor? _____

[13]

1000-watt

[14]

Which motor would supply more power, a 1000-watt motor or a 500-watt motor? _____

[14]

1000-watt

[15]

Energy is defined as the use of power for a given period of time. There must actually be a source for all forms of energy. The energy which a motor supplies must come from somewhere. In the case of an electric motor, the motor uses electricity to supply _____ to the equipment it is running.

[15]

energy

[111]

Since the fuse is between the supply source and the loads, all of the _____ required by the loads must first flow through the fuse.

[111] current

[112]

Because the total circuit current must first flow through the _____, the fuse must conduct the same amount of _____ as the conductor leading from the source to the loads.

[112] fuse

current

[113]

The fuse is actually just another piece of wire mounted inside a glass case. Therefore, we can use the _____ to protect the circuit conductor from carrying too much current.

[113] fuse

[114]

We determine how much current the fuse will allow the conductor to carry by designing the fuse to get hot and melt when a certain amount of _____ flows through it.

[114] current

[115]

A common-size circuit wire in your house is a No. 14 AWG (American Wire Gauge). The National Electrical Code specifies that the wire should not conduct more than 15 amp. Why, in your own words, is it necessary to limit the current that this wire should carry? _____.

[115] It will get hot; possible fire.

[16] Therefore, an electric motor does work for us and it, of course, uses _____ to develop its power in the form of <u>motion</u>.	[16] electricity
[17] We will see in future frames how we determine the amount of electricity used by a motor in developing power to operate or move equipment. It is this _____ which we pay for.	[17] electricity / use of power
[18] We have now discussed one form of energy, namely <u>motion</u>, such as the turning of motor shafts. We observed that the faster the same work is done, the [more/less] power is used.	[18] more
[19] Thus, when we connect a motor to an electric source and supply electric energy to the motor, we obtain _____, or movement, which is one form of energy.	[19] motion
[20] We have seen that _____ is one form of energy. The other common form is heat. We consider light to be in this category, since the light is a result of the heating of wires in a lamp. (See Panel V-A.)	[20] motion

[106] hotter

[106] Eventually, if the overcurrent becomes too great, the wire will become _____ and possibly even melt.

---

[107] current

[107] When the wire is this hot, it may easily start a fire if it is near combustible material. Note that although the wires in toasters, bulbs, and heaters become red hot due to _____ flowing in them, they are intentionally made to do so and in a controlled, safe manner.

---

[108] current

[108] The danger in allowing wires to become hot arises when the conditions are not controlled or desired, as is the case in the wires running through our house. If we are to prevent these circuit wires from overheating, we must limit the _____ flowing through them.

---

[109] current

[109] We use a fuse to limit the _____ in our house circuits.

---

[110] series

[110] If you examine a fuse, you will see that it contains a small wire. (See Panel V-E.) When you place the fuse into its socket, you are putting this wire in [series/parallel] with the circuit wire. (See Panel V-F.)

[21]

Just as is the case with motors and motion, the rate at which heat (or light) is developed is called _____. The continued use of this power for a given rate of time requires the supply of electric _____.

[21]

power

energy

[22]

Power, therefore, is a measure of the _____ at which motion takes place or heat (light) is developed for us. Energy is the continued use of this _____ for a given period of time.

[22]

rate / speed

power

[23]

When we turn on a light bulb, we get light (and heat). The larger the wattage of the bulb, the [more/less] light we get.

[23]

more

[24]

We saw before that when an electrical apparatus supplies energy to us in the form of motion, heat or light, it must itself be supplied with energy in the form of _____.

[24]

electricity

[25]

Light bulbs, as just mentioned, are rated in watts, as 60-watt, 100-watt, etc. Since _____ is the unit of measure for power, the rating of a bulb tells us how much electric _____ would be required to light the bulb.

[25]

watts

power

[101]

Wires are rated according to the amount of current they can carry without overheating. A wire rated at 10 amp can carry [more/less] than 10 amp of current without overheating.

less

[102]

If you will examine a bulb and a toaster that are off, you will find that they have something in common. The wires on the inside are very small compared with the size of the wires in the house electric circuit. Since the wires in these devices are small, you would expect them to conduct a _____ amount of current before becoming damaged.

small

[103]

However, although those wires in the devices are small, they are still connected across the full voltage in our house circuits and are forced, therefore, to carry more current than their size would normally allow. This will cause the wires to get _____.

hot

[104]

Think of a light bulb, say a 100-watt bulb. After it has been on for some time, it gets hot. This bulb is an electrical device; it has _____ and consumes _____.

resistance
power

[105]

You should now begin to see what happens to a wire that carries too much current. It gets _____.

hot

[26]

To light a 100-watt bulb requires 100 _____ of _____ _____.

[26]

watts

electric power

[27]

To get work or energy in the form of motion from an electric motor, we must supply electric _____ to the motor.

[27]

power

[28]

We have tried to establish an idea of what we mean when we say power. The faster we work, the [more/less] power is required.

[28]

more

[29]

If motor A does work faster than motor B can do the same work, motor ___ would have a higher wattage rating.

[29]

A

[30]

If an electric heater A heats a room faster than heater B can heat the same room, heater ___ has a higher _____ rating.

[30]

A

wattage / power

[96] Since the wires in our buildings are certain sizes (depending on how many devices we expect to connect), they will conduct only certain amounts of _____ before they become damaged.

[96] current / electricity

[97] Because the wires are limited by their size as to the amount of current they can conduct, we can add electrical devices to our circuit only until we start to exceed the amount of _____ which the wires in the circuit can safely carry.

[97] current

[98] In order to protect the wires in the circuit against too much current, we place a fuse in the circuit. Since the fuse is designed to melt at a certain amount of current, we can say that the _____ protects the wires from conducting too much current.

[98] fuse

[99] Fuses can be of many types, sizes, and shapes. (Refer to Panel V-D.)

[99]

[100] You should understand how a fuse works. The following frames will describe how excessive _____ will cause them to overheat.

[100] current

[31]

The more motion, heat, or light an electrical appliance supplies in a given period of time, the higher the _____ rating and the [more/less] the electric power required.

[31]

wattage

more

[32]

We mentioned before that when an electrical device develops energy in the form of motion or heat (light), it must be supplied with electric _____ for a given period of time.

[32]

power

[33]

It is the use of this electric energy which we pay for each month on our electric bill. Therefore, it is necessary to be able to measure this energy. This is done by an electric watt-hour meter, which measures the amount of power used during a given time period.

[33]

[34]

We must now see just how we can determine the exact amount of energy used. Since all electrical appliances furnish us either motion or heat and light, they all require electric _____, used over a certain period of time.

[34]

power

[35]

Any time we connect an electrical device to a voltage source we get current flow through it. At the same time, when we plug in our electrical device, we use electric _____, which is power used for a given time period.

[35]

energy

[91]

the same / equal

[91]

Therefore, our source voltages will always be considered constant and unchanging. This is generally true since our power systems supplying our homes are very large. If our supply voltage does not change, therefore, all of the loads we add in a parallel circuit will have _____ voltage.

[92]

current

[92]

Let us now take a closer look at power. We use power in our homes, schools, offices, etc., in three basic ways: To develop motion (motors), heat (iron, toaster), and light (bulbs). Regardless of the use for which we apply power, it is still equal in value to the product of the voltage drop of our electrical device times the _____ through the device.

[93]

current

[93]

Since we assume, then, that the supply voltage will not change regardless of the number of loads we add, it might seem that we could add an unlimited amount of electrical devices to our circuits. We know that this is not true. Since $P = IE$ and all electrical devices consume power, it must be that _____ is the limiting factor as to how many loads we can add to a parallel circuit.

[94]

decreases
current

[94]

As we add loads to a parallel circuit, the total circuit resistance _____, and the _____ flowing from the source increases.

[95]

Larger

[95]

As you saw earlier in this course, the amount of current that a wire of the same material can conduct depends in part on the size (thickness, diameter, etc.) of the wire. [Larger/Smaller] wires can carry more current.

[36] It appears, therefore, that there must be a relationship between the power required by a device and the voltage, current, and resistance of the device. This is the case. Power is related to a combination of the resistance of the device and <u>current</u> through it and also to a combination of the voltage supplied to a device and the _____ through it.	**[36]** current
[37] There is a relationship between the electric power used by a device, the voltage supplied to the device, and the _____ through it. (See Panel V-B.)	**[37]** current
[38] The relationship of the power requirement of a device to the voltage of the device is $\qquad P = IV.$ State this equation <u>in your own words</u>. _____	**[38]** Power equals current times voltage of appliance, amperes × volts.
[39] Since power equals the <u>current</u> flowing in a device multiplied by the <u>voltage</u> supplied to the device, we can find the electric power used by a device if we know its _____ and _____.	**[39]** current voltage
[40] The power needed to light a bulb can be found by the formula _____.	**[40]** $P = IV$

[86]
There is a relationship which we should consider when dealing with parallel circuits. Remember that P = IV (reference Frame 38). Every time we add an electrical device, it draws current and has a voltage drop. It therefore consumes _____ from our source.

[86] power

[87]
If we turn this around slightly, we can say that since all electrical devices require power to operate, they must all require current and voltage. In a parallel circuit we have already seen that each load has a voltage drop equal to the _____ voltage, and that the voltage drops of all loads are equal.

[87] source

[88]
In order for the load to obtain the power it needs to operate, it must have, in addition to the voltage already discussed, a _____ through it.

[88] current

[89]
Therefore, every time we add a load to a parallel circuit, we are increasing the needed power which the source must supply and, therefore, since the voltage remains the same, we must increase the supply _____.

[89] current

[90]
Under some conditions, we add too many loads to the parallel circuit and we overwork the supply source, causing its voltage to decrease. For our present purposes, however, we will consider that our source is large enough to handle all of the _____ we wish to connect to the circuit.

[90] loads

[41]

Since power has a unit of measure, namely watts, then the wattage of the light bulb can be found by the formula _____.

[41]

P (or watts) = IV

---

[42]

A bulb that needs 120 volts to force 3 amp of current through it would be a _____-watt bulb.

[42]

360

---

[43]

To return to the 100-watt light bulb, if we have a current of 1 amp, we must have a voltage of _____. Write the equation used to get this answer.

[43]

100 volts

P = I × V
Explanation: 100 = 1 × V;
               V = 100.

---

[44]

All electrical appliances, due to their internal wiring, have a certain amount of resistance built into them. An electric toaster is an extreme example of this. It has a large amount of _____ in it.

[44]

resistance

---

[45]

When we connect an electrical appliance to a voltage source, we have a complete circuit and obtain a flow of _____.

[45]

current

[81]   power

[81] This can be proved if you remember that an electrical device re-quires electric _____ before it can do work for us.

---

[82]   rate / speed

[82] Likewise, we saw that the electric power requirement of a device depends on the _____ at which it does the work and is related to the voltage and current required by the device.

---

[83]   V

[83] The power supplied to a device is equal to I^2R. If R is not known, use the equation $P = I \times \underline{\quad\quad}$.

---

[84]   current

[84] We saw before that we could write the power equation as $P = IE$ or IV (or $P = I^2R$). The same is again true. The power consumed by a load can be written $P = IV$ (or $P = I^2R$). This says that the power consumed by a load equals the _____ through the load multiplied by the voltage drop (V) of the load.

---

[85] I^2R IV / IE

[85] We have, therefore, arrived at two ways for determining the power used by an electrical device. One method relates power to current and resistance; the other relates power to current and voltage. Write each of these two equations.

$$P = \underline{\quad\quad} \qquad \text{or} \qquad P = \underline{\quad\quad}$$

[46]

We have seen that P = IV for electrical devices. In Ohm's law we saw that V = IR also. Let's substitute the expression IR for V in the first equation since IR is equal to V. We then have P = I × (I × R). This can be written P = I × I × R or

$$P = I^2R.$$

[46]

---

[47]

The specific relationship of power, current, and resistance is therefore P = _____, which says that power required by an electrical device is equal to the resistance of the device multiplied by the square of the _____ flowing through the device.

[47]

I^2R

current

---

[48]

By "squared" we mean that a number is multiplied by itself. For example,

3^2 means 3 × 3 = 9

10^2 means _____

I^2 means I × ___ = I^2 (The numerical answer depends on the value of I.)

[48]

10 × 10 = 100

I

---

[49]

Find the following values.

(1) 2^2 = ___

(2) 3^2 = ___

(3) I^2 = ___ if I = 5

[49]

(1) 4

(2) 9

(3) 25

---

[50]

If your answers agreed with those given for Frame 49, move to Frame 55. If you disagree with those given, go to Frame 51.

[50]

[76]

How much power must be supplied to the circuit below? _____

[76]
2350 watts

---

[77]

As we add loads to our circuit, therefore, the supply power must increase; as we unplug loads in a circuit, the required supply power will _____.

[77]

decrease

---

[78]

Even though the electricity is available to us whenever we want it, if we have zero appliances or loads connected to our circuit, we require _____ power from our supply.

[78]

zero

---

[79]

You can see that all of the power supplied to a circuit is used by the _____ that are connected to the circuit.

[79]

loads / appliances

---

[80]

Panel V-C shows that part of the total power supplied to the circuit is required by each _____. (Refer to Panel V-C, Parts (a) and (b).)

[80]

load / appliance

[51]	**[51]**
Somewhere we've confused you. Possibly you had as answers to Frame 49, $2^2 = 4$, $3^2 = 6$ and $5^2 = 10$. Remember that Frame 48 said, "by squared we mean a number is multiplied by itself." Thus, $2^2 = 2 \times 2 = 4$; $3^2 = 3 \times 3 = 9$; and $5^2 = 5 \times 5 = 25$. Therefore, $4^2 = \underline{\quad} \times \underline{\quad} = \underline{\quad}$.	$4^2 = 4 \times 4 = 16$
[52]	**[52]**
Now you see it. Fine! In the same sense, $I^2 = I \times I$ and $r^2 = r \times r$. Thus, in the first example, when $I = 6$, $I^2 = \underline{\quad} \times \underline{\quad} = \underline{\quad}$ (substituting the number 6 for I). If $r = 1$, $r^2 = \underline{\quad} \times \underline{\quad} = \underline{\quad}$.	$6 \times 6 = 36$ $1 \times 1 = 1$
[53]	**[53]**
Suppose that $I = 9$; then $I^2 = \underline{\quad}$ and, if $r = 7$, $r^2 = \underline{\quad}$.	81 49
[54]	**[54]**
If you had the correct answers to Frames 52 and 53, you have a good understanding of this idea and are ready to move ahead to Frame 55. If not, we suggest you review this by returning to Frame 51.	
[55]	**[55]**
Since $P = I^2R$, which is the same as writing $P = I \times I \times R$, we can always find the power required by an electrical device if we know the $\underline{\quad\quad}$ flowing through the device and the $\underline{\quad\quad\quad}$ of the device.	current resistance

[71]	[71]
We have seen that another way of finding the power used by a device is to measure the current through the device and the <u>voltage</u> across it and use the equation P = ___ × ___ .	I V
[72]	[72]
To summarize, then, the electric power used by a device can be determined by any two of these three quantities: _____ _____ , _____ , and _____ .	<u>current</u> through it <u>voltage</u> supplied to it <u>resistance</u> of it
[73]	[73]
Since we don't get something for nothing, our circuit can consume only the power that is supplied to it. Therefore, if we have a circuit containing two 100-watt bulbs, the power supplied to the circuit must be equal to ____ watts.	200
[74]	[74]
In a circuit, the supplied power should equal the sum of the individual power requirements of each load. The supply power must _____ as we add loads to the circuit.	increase
[75]	[75]
A circuit with a power supply of 1000 watts could supply the power requirement of ___ 100-watt light bulbs.	10

[56] If a length of house wire has a resistance of 1 ohm and 20 amp flowing through it, how much heat is developed in the wire? Heat = ____ watts.	[56] 400 Explanation: $P = I^2R =$ $20 \times 20 \times 1 = 400$ watts.
[57] If we have an iron which has a resistance of 10 ohms and requires 5 amp of current to heat, what is the power requirement of the iron? P = _____.	[57] 250 watts Explanation: $P = I^2R =$ $5 \times 5 \times 10 = 250$ watts.
[58] If we know the power rating of a device and its resistance, we can also compute the current from the same formula, $P = I^2R$. For example, a 400-watt light bulb with a resistance of 100 ohms would have a 2 amp of current flowing in it. $$P = I^2R, \qquad\qquad 4 = I^2,$$ $$400 = I^2 \times 100, \qquad\qquad I = 2.$$	[58]

[59]
 	
We solved the previous problem by finding the square root (symbol $\sqrt{}$) of 4. To find a square root, we ask ourselves, "What number multiplied by itself will give this answer? Hence, we ask ourselves in this case, "What number multiplied by itself will give 4? The answer is 2. What are the following square roots?
 	
    (1) $\sqrt{9}$ = ___   (3) $\sqrt{25}$ = ___
 	
    (2) $\sqrt{16}$ = ___   (4) $\sqrt{1}$ = ___
	[59]
	
	(1) 3
	
	(2) 4
	
	(3) 5
	
	(4) 1

[60] If you agreed with all the answers given for Frame 59, go ahead to Frame 66. If your answers disagreed with those given, go on to Frame 60.	[60]

[66]

If we have a 100-watt bulb with a resistance of 25 ohms, how much current will flow through it? _____

[66]

I = 2 amp

Explanation:

$I^2 = \dfrac{P}{R} = \dfrac{100}{25} = 4$, or

$I^2 = 4$. Since $2 \times 2 = 4$ and

$I^2 = I \times I$, then I = 2 amp.

---

[67]

If a 100-watt bulb has a resistance of 100 ohms, how much current is flowing through it? (Remember that $P = I^2R$.)

[67]

1 amp

Explanation:

$I^2 = \dfrac{100 \text{ watts}}{100 \text{ ohms}} = 1$, or

$I^2 = 1$. Since $1 \times 1 = 1$, then I must be 1 amp.

---

[68]

Two electrical appliances have the values given below. Solve for I and P.

(1) P = 250 watts   (2) I = 4 amp

R = 10 ohms   R = 25 ohms

I = _____ amp   P = _____ watts

[68]

(1) 5

(2) 400

Explanation: $I^2R = P$;

$4^2 \times 25 = 16 \times 25 = 400$.

---

[69]

A toaster of 1000 watts draws 10 amp of current. Find its resistance. R = _____ ohms.

[69]

10

Explanation: $P = I^2R$;

$1000 = 100R$, $10 = R$.

---

[70]

We have seen that the power required by an appliance is related to the _____ through the device and the resistance of the device by the equation P = _____.

[70]

current

I^2R

[61]

Since you're having some difficulty, let's look at our problem in this way. When we're asked, "What is the square root of 9, ($\sqrt{9}$)?" we are really being asked, "Can you think of the number which, when multiplied by itself, will give the result of 9? What is this number? ___

[61]

3

[62]

You see, here we are concerned with just the opposite idea to that in Frame 48. There we were considering the idea of squaring a number, like 3^2, and arriving at an answer of 9. Now we are really working backwards, starting out with the answer and trying to discover what the original number was that, multiplied by itself, would give the answer.

[62]

[63]

Thus, you should easily see that there is a direct relationship between squares and square roots. For example,

$2^2 = 2 \times 2 = 4$; therefore, $\sqrt{4} = 2$.

Complete the following statements.

$8^2 = 8 \times 8 = 64$; therefore, $\sqrt{64} =$ ___.

$4^2 =$ ___ $\times$ ___ $=$ ___; therefore $\sqrt{16} =$ ___.

[63]

8

$4 \times 4 = 16$

$\sqrt{16} = 4$

[64]

To review this idea let's get a few more square roots before we go on and apply it to solving a problem.

(1) $\sqrt{36} =$ ___   (2) $\sqrt{1} =$ ___   (3) $\sqrt{49} =$ ___

[64]

(1) 6

(2) 1

(3) 7

[65]

If all your answers to Frame 64 agreed with those given, you've probably got the idea now. Go ahead to Frame 66. If you're still having a problem, go back to Frame 61 for review.

[65]

You have completed the study of the fundamentals of electricity. The information that follows is designed to aid in your future work.

The first, Glossary, contains definitions of terms used in the electric utility business.

The second is an Index to the course that you have just completed. You should refer to this Index in the event that you need a detailed review of any major part of the course.

Glossary

AC (ALTERNATING CURRENT).   Current alternately turned on and off as a conductor in a generator makes one revolution or cycle.

AMMETER.   An instrument used to measure current.

APPLIANCE GROUND.   A conductor leading from the case of an appliance to a solid ground connection. Its purpose is to prevent the user of the appliance from receiving an electric shock from the case in the event that the insulation fails on the conductors connecting the appliance.

CIRCUIT.   Path of an electric current. It exists when a voltage source is connected to a resistance or load by conductors and is necessary in order for current to flow.

CIRCUIT BREAKER.   Part of a circuit which performs a function similar to that of a fuse in that it protects conductors from carrying too much current. It does not melt as does a fuse and can generally be "reset" by the pushing of a button on the circuit breaker itself. For large power applications, such as in a substation, these may be very large pieces of equipment.

CLOSED CIRCUIT.   A circuit that is complete and therefore allows current to flow.

CONDUCTOR.   A material or substance through which current can flow. Larger-sized conductors usually can carry more current than a small conductor of the same material.

CONTINUITY.   A word used to describe completeness of a circuit. A circuit in which current can flow is said to have continuity.

CURRENT.   Flow of electrons through a conductor, measured in amperes. Symbol for current is I.

CYCLE.   One complete revolution of a generator conductor through a magnetic field such that the induced voltage is twice "off."

DISTRIBUTION CIRCUIT.   Circuit containing the conductors which carry the power from a substation to the area where it is used by a customer.

1

ENERGY.   The quantity of electric power used in a given length of time, measured in watt-hours or kilowatt-hours. Example: A 100-watt bulb turned on for 10 hours would use the following amount of energy: 100 watts × 10 hours = 1000 watt-hours or 1.0 kilowatt hour.

ENTRANCE BOX OR FUSE PANEL.   In the home, a metal box in which the fuses or circuit breakers are located. This is where the house circuitry is connected to the wires coming out of the meter socket.

FUSE.   A device, consisting of a small meltable wire, placed in series with each building circuit. Designed to melt when current reaches a value that could damage the conductor through which it is flowing.

GENERATOR.   A device consisting of at least one conductor which is mechanically and continuously made to pass through a magnetic field, thus causing induced voltage in the conductor. It is the source for nearly all of our electric power.

GROUND WIRE.   A metallic conductor connected to ground by a ground stake, water pipe, etc., on one end and the device in question (appliance, etc.) grounded on the other.

INDUCED VOLTAGE.   A voltage which is made to appear on a conductor when the conductor is mechanically passed through a magnetic field.

INSULATOR.   A material designed to prevent flow of current as well as electrical shock to anyone near that source of voltage; that is, any nonconducting material — rubber, glass, etc.

KILOWATT-HOUR.   Amount of electric energy consumed by electrical devices in a measured amount of time; 1000 watts used for one hour.

KVA. 1000 VOLT-AMPERES.   The rating assigned to a distribution transformer; for example, 15 kva means 15,000 volt-amperes.

LOAD.   Power-consuming device, which contains resistance.

MAGNETIC FIELD.   A magnetic force around a conductor, which is caused by current through the conductor. It is similar to that caused by a bar magnet.

METER.   Device usually belonging to the power company; it measures the amount of kilowatt-hours used by the customer.

METER SOCKET.   Device usually furnished by the power company. It is attached to customer's home, and the meter inserted into it.

NEUTRAL (NEUTRAL WIRE).   The center wire in a 3-wire service. It is connected to ground and will have a 120-volt potential with reference to either hot wire in the service.

OHMMETER.   An instrument used to measure resistance. Can also be used as a continuity checker.

OPEN CIRCUIT.   A circuit that is incomplete or contains a broken wire, thus preventing current from flowing.

PARALLEL.   A circuit in which there are many parallel paths which the current can take. This type of circuit is used in house wiring.

POLES OF A MAGNET.   The surfaces on a magnet where the lines of force leave (north pole) and re-enter (south pole) the magnet.

POWER.   The conversion of electricity to work in the form of light, heat, and motion. It is measured in watts. Symbol for power is P. $P = EI$ or I^2R. Power factor must be absolutely accurate. (See "Power Factor.")

POWER FACTOR.   A number ranging from 0 to 1.0 which must be used in the computation of power required by an appliance or other load.

PRIMARY.   The incoming side of a transformer; the voltage (or sometimes current) which is being changed to a different value by the transformer. The primary side may be designated by P, pri, or primary.

PRIMARY CIRCUIT.   The conductors which carry the high voltage into residential areas.

RESISTANCE.   The opposition of a conductor to the flow of current. Measured in ohms. Has the symbol R.

RISER.   The part of the service entrance wiring which leads from the meter and connects to the service drop where it is fastened to the house.

SCHEMATIC.   A line diagram showing electric circuitry.

SECONDARY.   The outgoing side of a transformer. The new value of voltage (or sometimes current) to which the incoming voltage or current has been changed. The secondary side may be designated by S, sec, or secondary.

SECONDARY CIRCUIT.   Circuit containing the conductors which carry the low voltage (120/240) into residential area from the transformers to the customer's home.

SERIES CIRCUIT.   A circuit in which the current has only one path to take.

SERVICE DROP.   The lines connecting the power company's distribution system and the home. These service drops are usually 3-wire and occasionally 2-wire.

SERVICE DROP, THREE-WIRE. A service drop which contains a neutral and 2 "hot" wires and serves the customer with 120/240 volts.

SERVICE DROP, TWO-WIRE.   A service drop which contains a neutral and a "hot" wire and serves the customer with 120 volts only.

SERVICE ENTRANCE.   Unit which includes the riser, meter installation, and fuse box.

SHORT CIRCUIT.   A circuit in which resistance has been reduced so far that it allows a dangerously high current to flow.   Fire may result.

SUBSTATION.   An enclosure which contains transformers, circuit breakers, fuses, switches, and other electrical apparatus.   A place where voltage levels are changed by transformers and where circuits (conductor, lines, etc.) terminate and/or originate.

SWITCH.   A mechanical means of opening or closing a circuit.

TRANSFORMER.   A device consisting of a set of windings used to change voltage or current from one value to another.

TRANSFORMER FUSE.   A fuse mounted near a transformer, protecting it the same way the fuse in our house protects our wiring.

TRANSMISSION CIRCUIT.   The conductors which carry the power from a generating station to the substation.

VA (VOLT-AMPERES).   The rating that is usually assigned to small-appliance transformers.

VOLTAGE.   Potential difference causing current in a conductor. It is measured in volts, and the symbol for voltage can be V or E.

VOLTMETER.   An instrument used to measure voltage (potential difference).

WATT.   The unit in which power is measured.   See "Power."

WATT-HOUR.   The unit for measuring energy.   See "Kilowatt-hour."

WATT-HOUR METER.   An instrument used to measure energy consumed by connected loads.   Energy equals power × length of time power is used.

WATTMETER.   An instrument used to measure power consumed by connected loads.

Index

Post-test

To help you determine the value of the programed learning course, Fundamentals of Electricity, Volume 1: Basic Principles, please answer the following questions to the best of your ability.

1. The amount of electric current flowing in a circuit is measured in _____ .

2. Electrical components can be connected in circuits in two ways. These two kinds of circuits are called _____ and _____ .

3. The formula which expresses the relationship between voltage, current, and resistance in an electric circuit is ___ = _____ and is called _____ _____ .

4. The amount of resistance in an electric circuit is measured in _____ .

5. Two conductors are of equal length but one has twice the diameter of the other. The [larger/smaller] one would have the most resistance. The [larger/smaller] would carry the most current.

6. A device that contains a metal strip with a low melting point, installed in a circuit to protect against overloading, is called a _____ .

7. _____ is the unit of measure for the pressure which causes current to flow through resistance in an electric circuit.

8. Power used in an electrical device is measured in _____ and and is computed by the formula _____ .

9. The drawing below illustrates a simple _____ circuit.

120 volts

24 ohms

10. In the circuit above, a current of ____ amp will flow.

11. The term kilo, sometimes used with electrical terms, means _____ . A kilowatt, therefore, is equal to _____ watts.

12. When a 2200-watt load is connected in a 120-volt circuit, will a 15-amp fuse hold or will it blow? [Hold/Blow.] Why? _____

13. What is the principal reason that some homes have a three-wire entrance service rather than a two-wire service? _____

1

14. In every electric circuit, the current flows from the [positive/negative] terminal to the [positive/negative] terminal.

15. A [copper/glass] rod is the best conductor of electricity.

16. A current of ___ amp will flow in the circuit below.

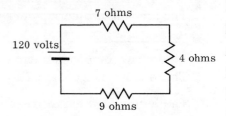

17. A No. 14 conductor [will/will not] safely carry a 100-amp load. Why? _____

18. The drawing below indicates a _____ circuit.

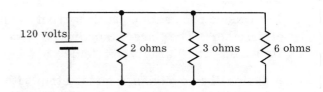

19. The total resistance in the circuit in Problem 18 is ___ ohm(s).

20. A circuit carrying 4 amp of current flowing through a 25-ohm resistance develops _____ watts of power.

21. If an additional 8-ohm resistance was added to the circuit in Problem 18, the total current for the circuit would [increase/decrease].

22. In some modern electric water heaters, 240 volts are supplied to a 4800-watt element. How much current will this element draw? _____

23. A transformer has 100 turns in the primary windings and 20 turns in the secondary. If the primary voltage is 120 volts, the secondary voltage would be ____ volts.

24. The transformer described in Problem 23 would be a [step-up/step-down] transformer.

25. Current in a conductor causes a certain amount of _____ _____, which may result in low voltage being supplied to the load.

26. With a voltage generated at 60 cycles per second, how many times will the electricity be "off" in a five-second interval? _____

27. What happens to a wire that is forced to carry an amount of current in excess of its rated capacity? _____ _____

28. The normal circuit wiring in a house is an example of a [series/parallel] circuit.

29. A load that has 40-ohm resistance and is rated at 1000 watts would have _____ amp of current flowing through it.

30. A motor rated at 120 volts, if connected to a 240-volt outlet, would _____ _____.

31. How much does it cost to operate a 1000-watt iron for 1/2 hour at a rate of 2 cents per kWh? _____

32. In some electrical appliances (washers and dryers, for example) a wire is often connected from the case of the appliance to a cold-water pipe. This wire is called a _____ wire. What is its purpose? _____

33. If you noticed in your home that your lights were dim, motors were running slowly, etc., you would assume that you had a condition of _____ _____.

Answers to Post-test

1. amperes (amp)

2. series
   parallel (either order)

3. $E = IR$, $I = \dfrac{E}{R}$, $R = \dfrac{E}{I}$
   (any one of these)
   Ohm's law

4. ohms

5. smaller/larger

6. fuse

7. Volt

8. watts
   $P = EI$ or $P = I^2R$

9. series

10. 5

11. 1000
    1000

12. Blow.
    15-amp fuse good for only
    1800 watts.

13. They need 120/240 volts.

14. positive
    negative

15. copper

16. 6

17. will not
    No. 14 rated at 15 amp.

18. parallel

19. 1

20. 400

21. increase

22. 20 amp.

23. 24

24. step-down

25. voltage drop

26. 600.

27. Heats up.

28. parallel

29. 5

30. burn out

31. 1 cent.

32. ground
    To prevent electric shock.

33. low voltage

Panel I-A
STRUCTURE OF ATOM

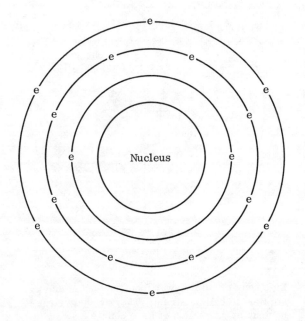

Panel I-D
STRUCTURE OF CHARGES OF NEGATIVELY CHARGED ATOM

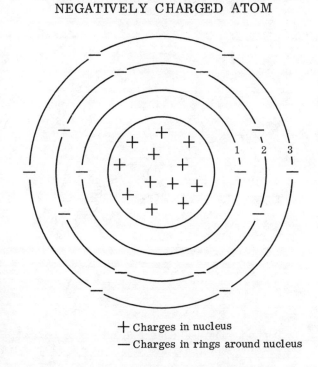

+ Charges in nucleus
− Charges in rings around nucleus

Panel I-B
ATTRACTION OF CHARGES

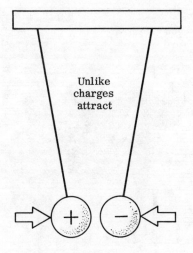

Iron balls with charges as shown will attract each other.

Panel I-C
REPULSION OF CHARGES

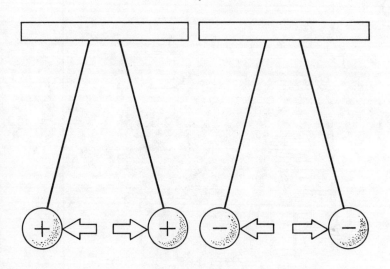

Like positive charges repel each other.

Like negative charges repel each other.

Panel I-E

FORCE ON BALL DUE TO MOTION

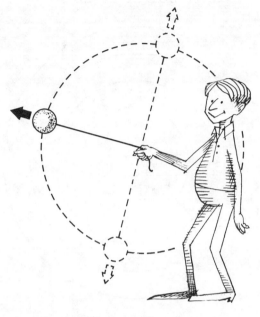

Due to motion, ball tries to fly off
end of string.

(a)

FORCE ON ELECTRON DUE TO MOTION

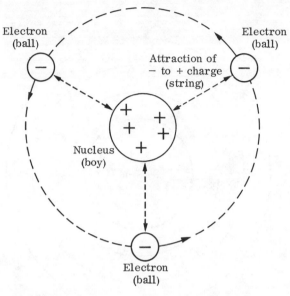

Electrons try to fly away from nucleus of atom
similar to ball on string. They are held close to
nucleus by attraction of its + charge.

(b)

Panel I-F

CONDITION OF PHYSICAL CHARGES
PRIOR TO LIGHTNING

Earth has deficiency of electrons (+)

Panel I-G

LIGHTNING CAUSED BY
IMBALANCE OF CHARGES

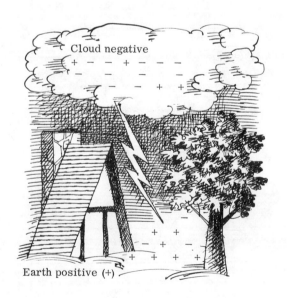

Electrons flow to earth through
lightning path.

Panel I-H

CONDITION OF PHYSICAL CHARGES
AT COMPLETION OF LIGHTNING

No flow of electrons since earth and cloud
have the same charge.

Panel I-I

FLOW OF ELECTRONS IN WIRE

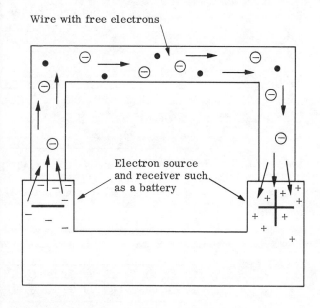

⊖ Free electrons
● Atoms

Panel I-J

ELECTRICAL TERMS COMMONLY USED

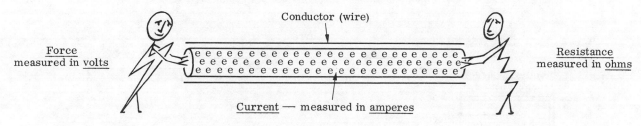

Electric force tries to push current through the conductor and resistance tries to
oppose the movement of this current. We can also say that volts try to push amperes
through the conductor and ohms try to oppose the movement of this current.

Battery serves as source of
electric force to push current
through the conductor.

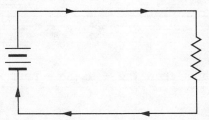

Resistance tries to oppose the
movement of current through
the conductor.

Current is pushed through the circuit and
the resistance by the voltage.

Panel I-K

RESISTANCE OF WIRES DUE TO SIZE

(It is understood that both wires carry the
same current; that is, they have the same
number of electrons flowing through them.)

(1)

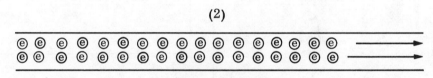

LARGE WIRE

Electrons have freedom of movement.

(2)

SMALL WIRE

Electrons are squeezed.

Panel I-L

DIRECTION OF CURRENT

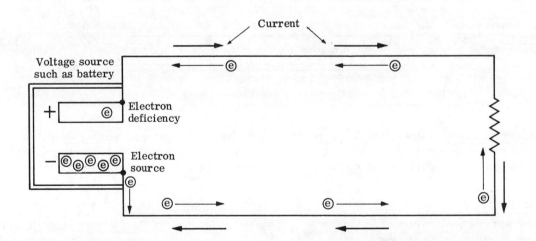

Electrons, ⓔ, try to move from the − to the + part of the circuit or battery.

The current is in the opposite direction to electron flow or, therefore, in the
direction of + to −.

Panel II-A

COMPONENTS OF A CIRCUIT

Battery (source),
conductors (wire),
and resistance (bulb) = circuit

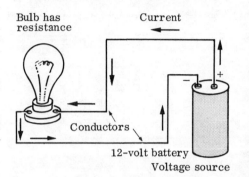

Bulb has resistance

Current

Conductors

12-volt battery
Voltage source

Complete circuit which
will allow a current

Panel II-B

CONDUCTIVE AND NONCONDUCTIVE MATERIALS

Conductors

Silver	Bronze
Copper	Solder
Aluminum	Impure water
Steel	Carbon
Iron	Earth

Nonconductors

Rubber	Porcelain
Glass	Wood

Panel II-D

FORMS OF OHM'S LAW

$$I = \frac{E}{R},$$

to solve for current;

$$E = IR,$$

to solve for voltage;

$$R = \frac{E}{I},$$

to solve for resistance.

Panel II-C

ELECTRICAL SYMBOLS

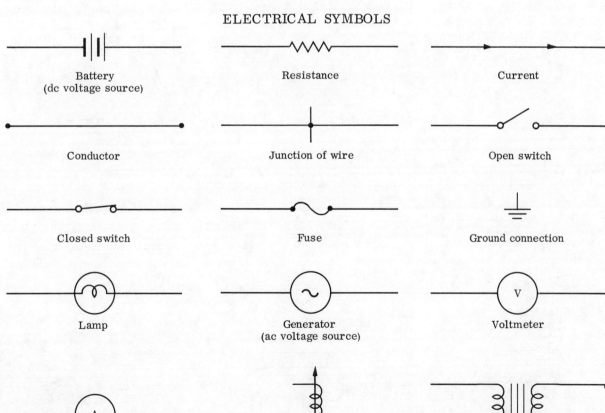

Battery
(dc voltage source)

Resistance

Current

Conductor

Junction of wire

Open switch

Closed switch

Fuse

Ground connection

Lamp

Generator
(ac voltage source)

Voltmeter

Ammeter

Solenoid

Transformer

Panel III-A
CURRENT IN A SERIES CIRCUIT

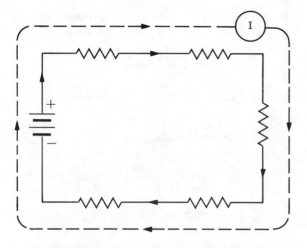

Current can flow only in one path
in series circuit.

Panel III-B
CURRENT AND VOLTAGE

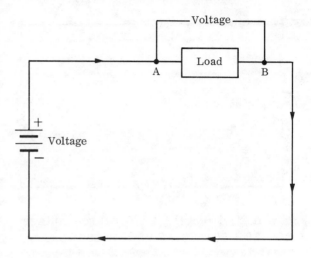

Voltage can exist only across loads
and at source.

Panel III-C
WHERE VOLTAGE EXISTS IN CIRCUIT

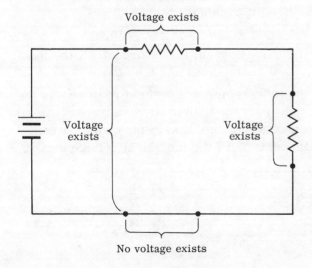

Voltage exists only when there is a load or
resistance between the points checked.

Panel III-D
VOLTAGE IN A CIRCUIT

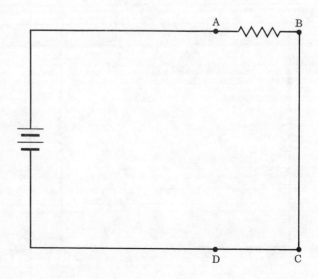

Panel III-E

CIRCUIT VOLTAGES

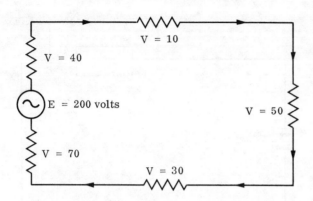

Voltage drops (V)   =   source voltage
40 + 10 + 50 + 30 + 70 =   200 volts

Note that in order for the circuit current to flow through each load, the load must have a pressure (voltage) across it. Each load uses its share of the source voltage and the total of these load voltages must equal the source voltage.

Panel III-F

VOLTAGE DROP

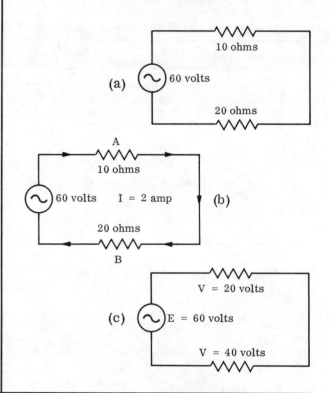

Panel III-G

SERIES CIRCUIT

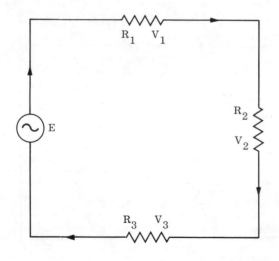

Panel III-H

REVIEW OF SERIES CIRCUITS

As a final reminder of the characteristics of a series circuit, remember that

1. all loads are connected end to end,

2. the same current flows through all the individual resistances and the source,

3. the total resistance of the circuit equals the sum of all the individual resistances (loads); the total voltage drop of the circuit equals the sum of all the individual voltage drops.

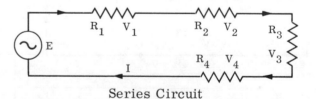

Series Circuit

E = total circuit voltage = $V_1 + V_2 + V_3 + V_4$.

$R_{Total} = R_1 + R_2 + R_3 + R_4$.

Same current (I) is in all the loads and the source.

Panel IV-A
COMPARISON OF CIRCUITS

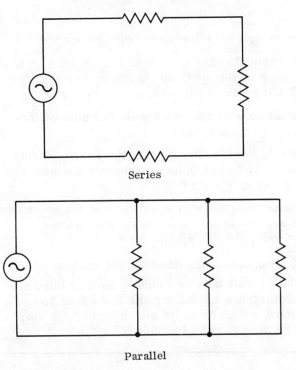

Series

Parallel

Panel IV-B
PARALLEL CIRCUIT
Current split at junction of conductors

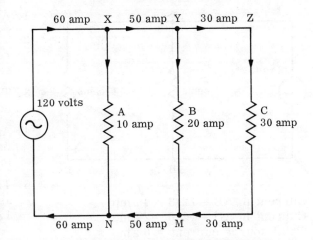

Current flow is in the direction of the arrows as shown. Conductors nearer the source carry the most current in the circuit. [NOTE: A junction in an electric circuit exists where current either splits or combines.]

Panel IV-C
PARALLEL CIRCUIT

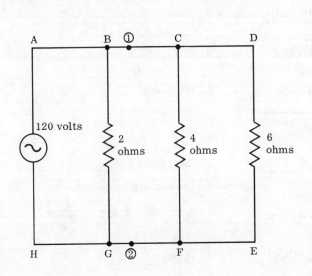

Panel IV-D
PARALLEL CIRCUITS
(Examples of household wiring)

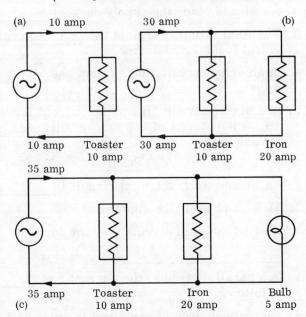

Note that as more loads are added, the source current increases.

Panel IV-E

SUMMARY OF PARALLEL CIRCUITS

Part A

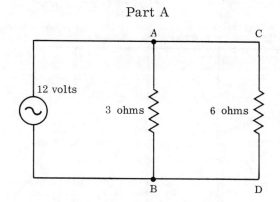

Current in AB = 12/3 = 4 amp.
Current in AD = 12/6 = 2 amp.
Total source current = 6 amp.

$$\frac{1}{R_T} = \frac{1}{3} + \frac{1}{6} = \frac{2}{6} + \frac{1}{6} = \frac{3}{6} = \frac{1}{2};$$

hence, $R_T = 2$, $I = \frac{E}{R} = \frac{12}{2} = 6$ amp.

Part B

1. All loads will have the same voltage drop.

2. The current through each load will depend on the resistance of the load and therefore is not necessarily the same in all loads.

3. The total source current equals the sum of the currents in each load.

4. As we add loads to the circuit, the source current increases and the conductors nearer the source must carry more current.

5. The total resistance in a parallel circuit decreases as we add loads and is found by the formula
$1/R_T = 1/R_1 + 1/R_2 + 1/R_3 + \cdots$

6. The total current flowing from the source in a parallel circuit may be found by determining the total resistance of the circuit and using the equation $I = E/R$ or by adding together the currents flowing in each individual load.

Panel IV-F

SUMMARY: SERIES AND PARALLEL CIRCUITS

Part A
Series

1. As we add load, total circuit resistance R _____.

2. As we add load, the supply current _____.

3. The total circuit R equals the _____ of all the individual loads or resistances.

4. In a series circuit, the current has _____ path(s) to follow.

5. In a series circuit, the same current flows through all loads but part of the source voltage is _____ at each load.

Parallel

1. As we add load, the total circuit R _____.

2. As we add load, the supply current _____.

3. We find the total circuit R by the formula _____

_____.

4. In a parallel circuit, the current has _____ path(s) to follow.

5. The voltage drop of each load in a parallel circuit equals the _____ voltage, but the _____ supplied by the source splits and part flows through each load.

Part B
Answers

1. increases

2. decreases

3. sum

4. one

5. dropped

1. decreases

2. increases

3. $\frac{1}{R_T} = \frac{1}{R_1} + \frac{1}{R_2} + \frac{1}{R_3} + \cdots$

4. many

5. source

current

Panel V-A

POWER DEVELOPING LIGHT AND HEAT

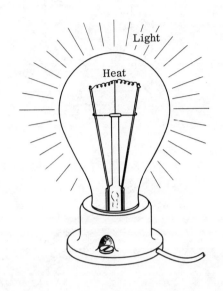

Panel V-B

WHEN POWER IS USED

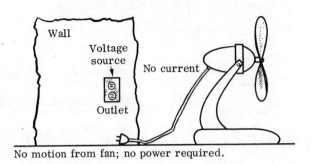

No motion from fan; no power required.

WHEN WE PLUG IN THE FAN

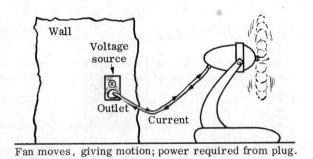

Fan moves, giving motion; power required from plug.

Panel V-C

FINDING TOTAL WATTAGE

(a)

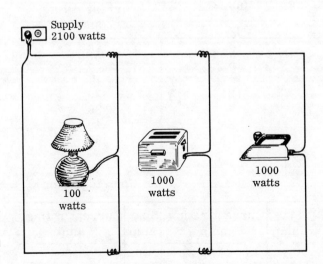

Watts used   watts supplied
100 + 1000 + 1000 =   2100 watts

(b)

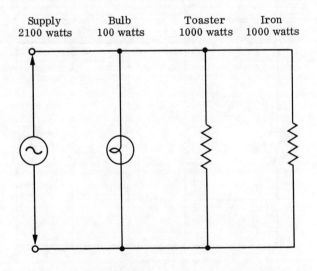

Watts used   watts supplied
100 + 1000 + 1000 =   2100 watts

Panel V–D
FUSES

(a)
Household fuses

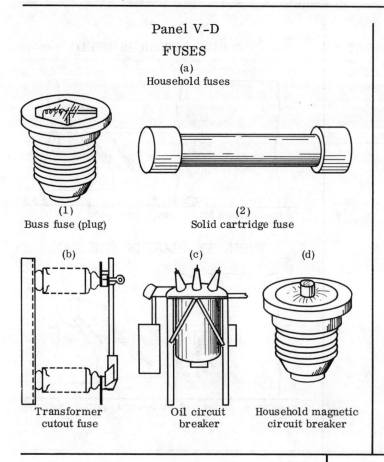

(1)
Buss fuse (plug)

(2)
Solid cartridge fuse

(b)   (c)   (d)

Transformer
cutout fuse

Oil circuit
breaker

Household magnetic
circuit breaker

Panel V–E
FUSE DESIGN

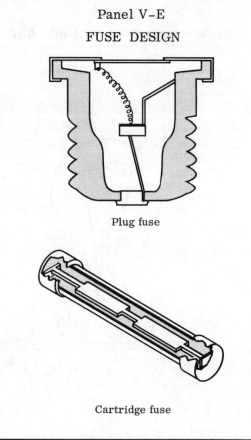

Plug fuse

Cartridge fuse

Panel V–F
FUSE APPLICATION

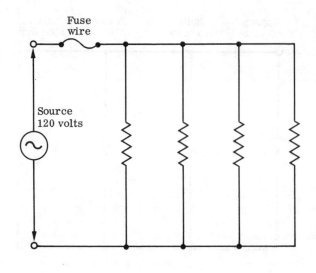

Panel V–G
WIRE AND FUSE SIZES: PARTIAL LIST
Household circuits

Wire

Size	Current rating, amp
18	6
16	8
14	15
12	20
10	30
8	40
6	55

Fuses

Plug type		Cartridge type	
Size, amp	Current rating, amp	Size, amp	Current rating, amp
10	10	40	40
15	15	60	60
20	20		
25	25		
30	30		

PANEL V-H

SUMMARY OF POWER CONCEPTS

General Information

1. Power is the rate at which work is done. Therefore, large amounts of power are required to do large amounts of work.

2. Work, in electrical terms, is related to two quantities. These are _motion_, as in motors, and _heat_, as in toasters. We consider light to be in the same category as heat, since it generally results from heat in some form.

3. The unit for measuring power is watts or kilowatts. Thus for example, a 100-watt bulb requires 100 watts of electrical power.

Power Equations

1. The power supplied to a device equals the power required by the device.[*]

2. The power required by a device may be determined in two ways:

   $P = $ (current through device $)^2 \times$ (resistance of device) $= I^2 \times R$,

   $P = $ (current through device) $\times$ (voltage drop of device) $= I \times V$.[*]

3. The power supplied to a circuit is equal to the supply current multiplied by the supply voltage:

   P (supplied) $= IE$.

4. The power supplied to a circuit equals the sum of the individual load power requirements:

   $$P_{supply} = P_{load\ 1} + P_{load\ 2} + \cdots$$

Fuses

1. Since overcurrent can cause wires to overheat, the current must be limited to safe values.

2. All loads require current which can come only from the same power source.

3. The conductor leading from the source to the first load must carry the total current supplied and is protected against overheating by a fuse.

4. Current above the rating of a fuse will cause the fuse to melt open and thus interrupt the current flow in the conductor. This prevents damage to the conductor.

---

[*]When we deal with power on ac-circuits, the power factor must be considered. This will not be discussed in detail in this course, but ac-power should not be computed without the use of the circuit power factor.

Panel VI-A
CONDUCTOR VOLTAGE DROP

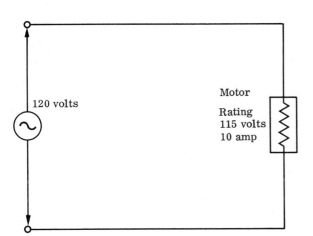

120 volts

Motor

Rating
115 volts
10 amp

Panel VI-B
CONDUCTOR VOLTAGE DROP

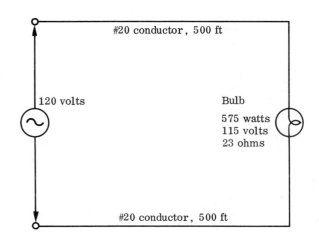

#20 conductor, 500 ft

120 volts

Bulb

575 watts
115 volts
23 ohms

#20 conductor, 500 ft

Panel VI-C
CONDUCTOR VOLTAGE DROP

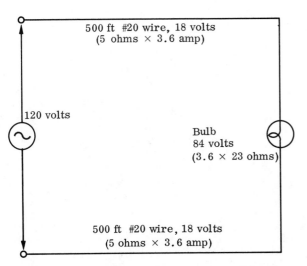

500 ft #20 wire, 18 volts
(5 ohms × 3.6 amp)

120 volts

Bulb
84 volts
(3.6 × 23 ohms)

500 ft #20 wire, 18 volts
(5 ohms × 3.6 amp)

120 volts from source = 18 + 84 + 18 = 120 volts

Panel VI-D
CONDUCTOR VOLTAGE DROP

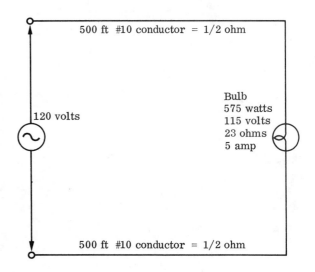

500 ft #10 conductor = 1/2 ohm

120 volts

Bulb
575 watts
115 volts
23 ohms
5 amp

500 ft #10 conductor = 1/2 ohm

Panel VII-A
MAGNETIC FIELDS

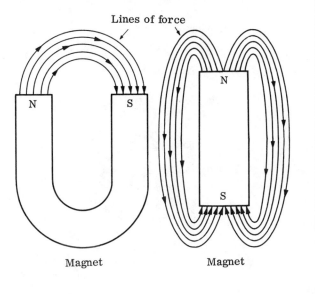

Lines of force

N   S

N

S

Magnet   Magnet

Panel VII-B
FORCING CURRENT TO FLOW BY MAGNETIC FIELDS

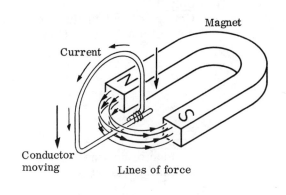

Magnet

Current

N

S

Conductor moving

Lines of force

Panel VII-C

VOLTAGE INDUCTION
(Magnet moving past conductor)

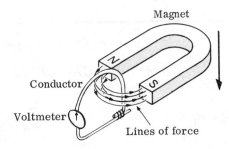

Magnet

Conductor

N

S

Voltmeter

Lines of force

<u>Step 1</u>
Conductor not in lines of force;
no voltmeter deflection

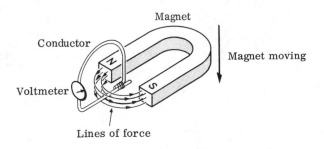

Magnet

Conductor

N

S

Voltmeter

Magnet moving

Lines of force

<u>Step 2</u>
Conductor in lines of force;
voltmeter deflects

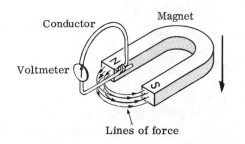

Magnet

Conductor

N

S

Voltmeter

Lines of force

<u>Step 3</u>
Conductor not in lines of force;
no voltmeter deflection

Panel VII-D

VOLTAGE INDUCTION
(Conductor moving past magnet)

Step 1
Conductor not in lines of force;
no voltmeter deflection

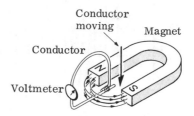

Step 2
Conductor in lines of force;
voltmeter deflects

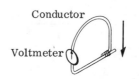

Step 3
Conductor not in lines of force;
no voltmeter deflection

Panel VII-E

MECHANICAL INDUCTION OF VOLTAGES

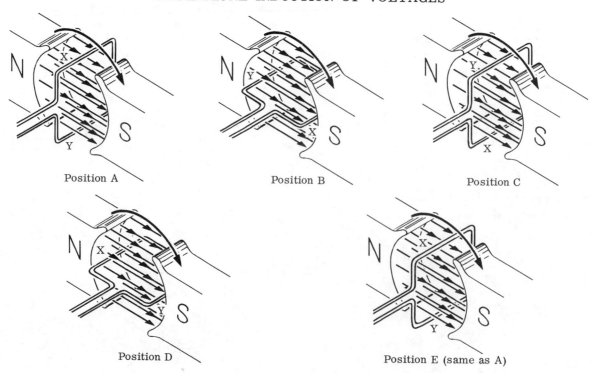

Position A   Position B   Position C

Position D   Position E (same as A)

Panel VII-F
USE OF TRANSFORMER

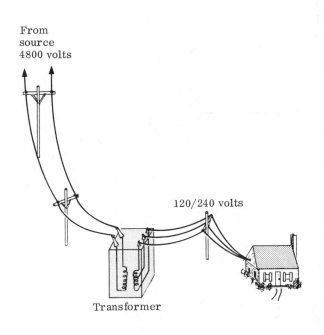

From
source
4800 volts

120/240 volts

Transformer

Panel VII-G
CONSTRUCTION OF TRANSFORMER

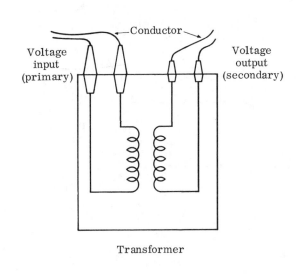

Conductor

Voltage
input
(primary)

Voltage
output
(secondary)

Transformer

Panel VII-H
TRANSFORMER TYPES

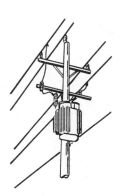

Pole transformer

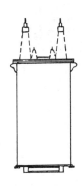

Single-phase power

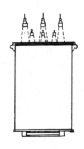

Three-phase power

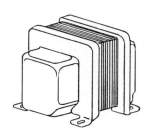

Appliance type

Panel VII-I

MAGNETIC FIELD DUE TO CURRENT

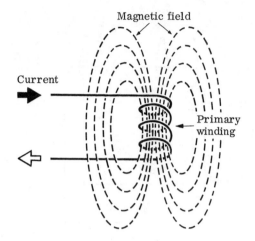

Panel VII-J

MAGNETIC LINKAGE OF WINDINGS
IN TRANSFORMER

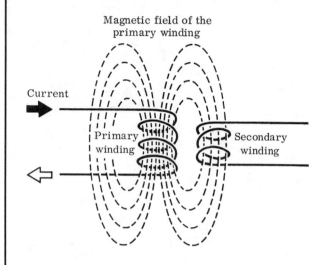

Panel VIII-A

A ONE-LINE DIAGRAM SHOWING A TYPICAL METHOD OF BRINGING GENERATED ELECTRICITY TO THE HOME

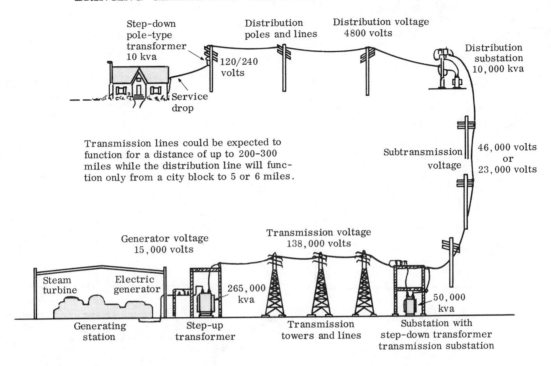

Step-down pole-type transformer 10 kva

120/240 volts

Service drop

Distribution poles and lines

Distribution voltage 4800 volts

Distribution substation 10,000 kva

Subtransmission voltage   46,000 volts or 23,000 volts

Transmission lines could be expected to function for a distance of up to 200-300 miles while the distribution line will function only from a city block to 5 or 6 miles.

Generator voltage 15,000 volts

Transmission voltage 138,000 volts

Steam turbine   Electric generator

265,000 kva

50,000 kva

Generating station

Step-up transformer

Transmission towers and lines

Substation with step-down transformer transmission substation

---

Panel VIII-B

SECONDARY DISTRIBUTION SYSTEM

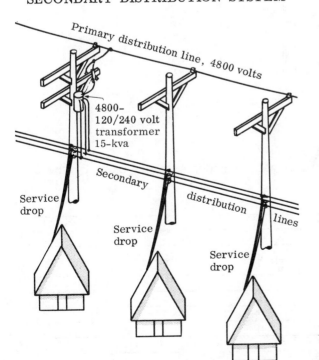

Primary distribution line, 4800 volts

4800- 120/240 volt transformer 15-kva

Secondary

distribution

lines

Service drop

Service drop

Service drop

Panel VIII-C

INTERNAL VIEW OF TRANSFORMER

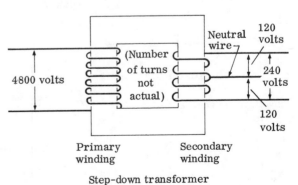

4800 volts

(Number of turns not actual)

Neutral wire

120 volts

240 volts

120 volts

Primary winding

Secondary winding

Step-down transformer
(4800-volt primary, 120/240-volt secondary)

Panel VIII-D
EXTERIOR VIEW OF TRANSFORMER

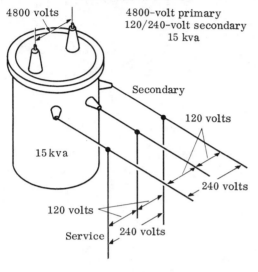

4800 volts

4800-volt primary
120/240-volt secondary
15 kva

Secondary

120 volts

15 kva

240 volts

120 volts

Service   240 volts

Panel VIII-F
COMPUTATION OF WATT-HOURS

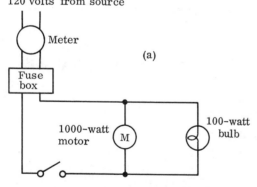

120 volts from source

Meter

(a)

Fuse box

1000-watt motor   M   100-watt bulb

Switch closed for 2 hours

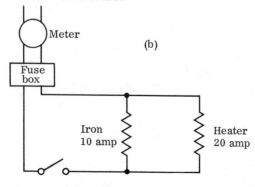

120 volts from source

Meter

(b)

Fuse box

Iron 10 amp   Heater 20 amp

Switch closed for 4 hours

Panel VIII-E
SERVICE ENTRANCE

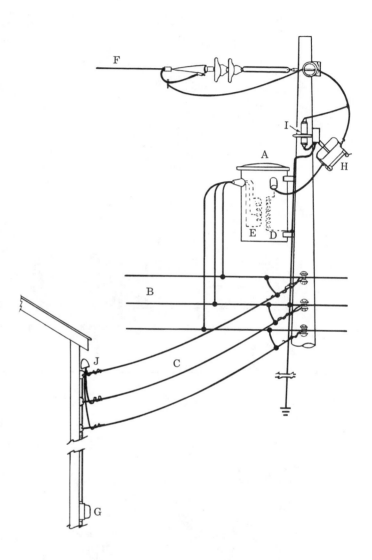

F

I

A

H

E   D

B

J   C

G

Complete with Correct Answers from Frame 35.

A. _____
B. _____
C. _____
D. _____
E. _____
F. _____
G. _____
H. _____
I. _____
J. _____

Panel VIII-G

SERVICE ENTRANCE VOLTAGES

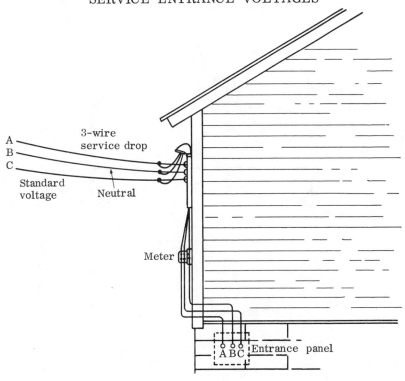

Panel VIII-H

ENTRANCE PANEL AND HOUSEHOLD CIRCUITS

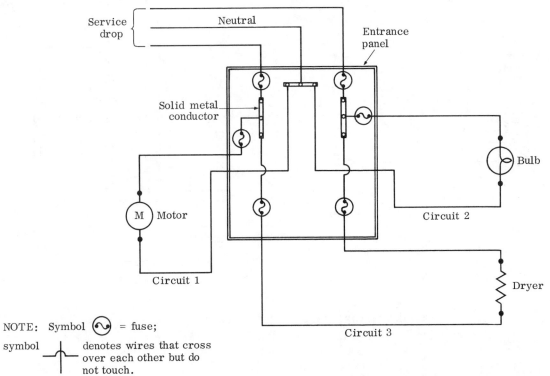

Panel VIII-I
APPLIANCE GROUNDING

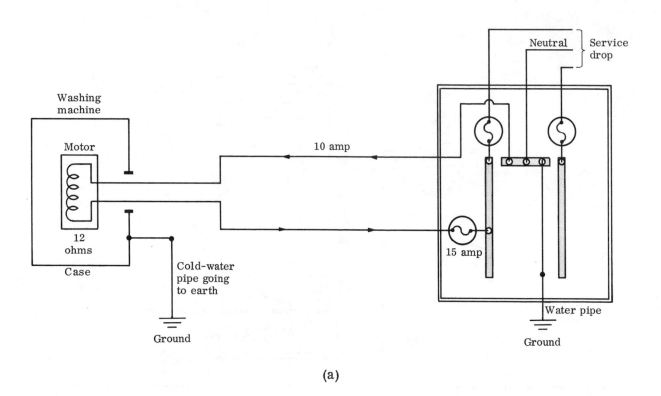

(a)

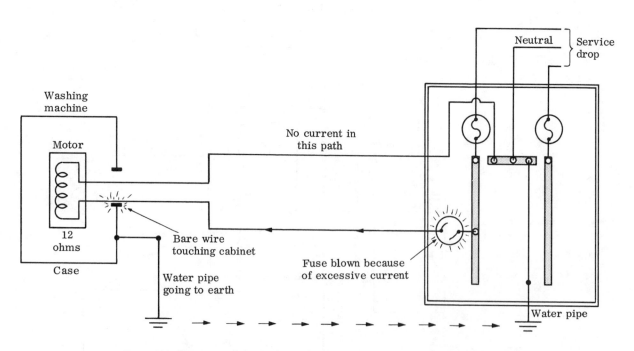

Practically zero resistance in this path;   therefore excess
current will blow a 15-amp fuse.

(b)

Panel VIII-J

APPLIANCE GROUNDING

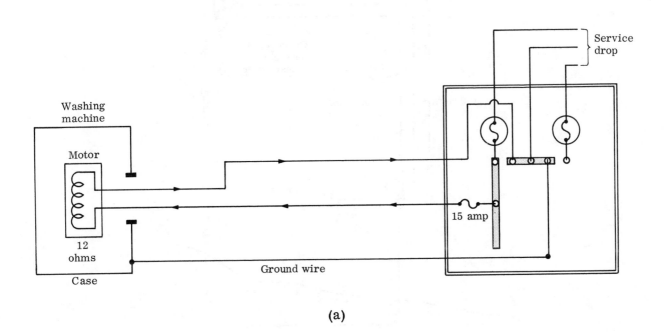

(a)

(b)

The ground wire has practically zero resistance, thus allowing excess current flow, which in turn blows the fuse.

Panel VIII-K

VOLTAGES OF 3-WIRE SERVICE WITH NEUTRAL

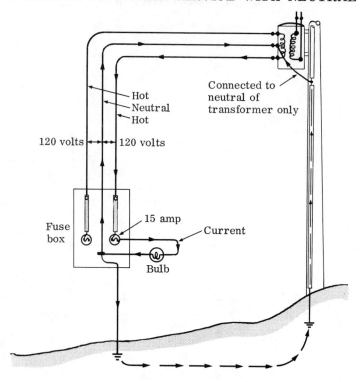

Panel VIII-L

SYSTEM NEUTRAL GROUND

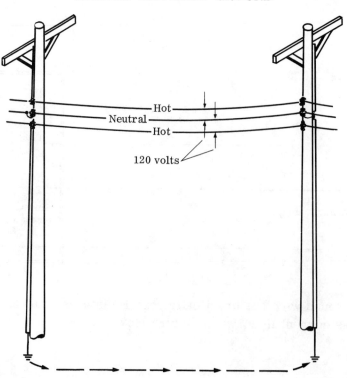

PANEL VIII-S1. SUMMARY

1. Electricity

Electricity, or current, is the flow of electrons (charged particles); it can exist only if there is a complete circuit. That is, there must be a source of potential difference and a conductor through which the electrons can flow. Ohm's law, which states the relationship between voltage, current and resistance in a circuit, can be given as follows: $E = IR$, with variations.

2. Parallel Circuits

(a) The type found in our household wiring.

(b) Supply current increases as more loads are added to the circuit.

(c) Example

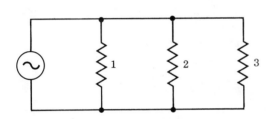

(d) The voltage across all loads is the same and equal to the supply voltage.

(e) The current through each load depends on the resistance of that load.

(f) Formulas:

$$\frac{1}{R_T} = \frac{1}{R_1} + \frac{1}{R_2} + \frac{1}{R_3} + \cdots,$$

$$I(\text{supply}) = I_1 + I_2 + I_3 + \cdots,$$

$$I(\text{supply}) = \frac{E}{R_T}.$$

3. Series Circuits

(a) Supply current decreases as more loads are added to the circuit.

(b) Example

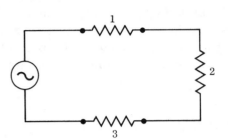

(c) The current through all the loads is the same and equal to the supply current.

(d) The voltage drop across each load depends on the resistance of the load.

(e) Formulas:

$$R_T = R_1 + R_2 + R_3 + \cdots,$$

$$E(\text{supply}) = V_1 + V_2 + V_3 + \cdots,$$

$$I(\text{supply}) = \frac{E}{R}.$$

4. Power

(a) The electric power used by any device, including a conductor, may be found by the following formulas:

$$P = I^2 R,$$

$$P = IE \quad \text{or} \quad IV.$$

(b) The source for a circuit will supply only as much power as the circuit requires.

(c) The loads on a circuit can use only as much power as the source can supply. Loads cannot make their own power supply.

(d) Energy is in the form of heat, light, and motion.

5. Fuse

(a) A fuse is a device connected in series with a household circuit.

PANEL VIII-S2. SUMMARY (cont.)

(b) The amount of current a circuit can carry is limited by a fuse, which melts when it is forced to carry too much current, and thus opens the circuit. This protects circuit conductors from overheating.

6. Magnetism

(a) When current of varying magnitudes flows through a wire, a magnetic field is set up.
(b) This magnetic field can cause a piece of metal to become a magnet with north and south poles.

(c) This magnetic field can force or induce a voltage to be set up in adjacent wires.

7. Transformers

(a) A transformer utilizes this concept of magnetism in changing voltages from a company's large transmission lines to voltages we use in our homes.

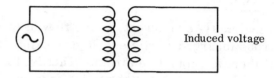

Induced voltage

(b) One side of a transformer winding is called the primary; the opposite side is called the secondary.
(c) Each side of the transformer has a winding of several loops of wire. The ratio of the primary to the secondary turns or loops determines how much the voltage is changed.
(d) Example

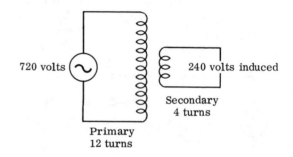

720 volts   240 volts induced

Secondary
4 turns

Primary
12 turns

Step-down transformer

Ratio = 12/4 = 3; thus 720/3 = 240 volts.